W9-ADT-979

HONEYBEE DEMOCRACY

HONEYBEE DEMOCRACY

THOMAS D. SEELEY

PRINCETON UNIVERSITY PRESS
PRINCETON AND OXFORD

Copyright © 2010 by Princeton University Press

Published by Princeton University Press, 41 William Street,
Princeton, New Jersey 08540
In the United Kingdom: Princeton University Press, 6 Oxford Street,
Woodstock, Oxfordshire OX20 1TW
press.princeton.edu

All Rights Reserved

Library of Congress Cataloging-in-Publication Data

Seeley, Thomas D.
Honeybee democracy / Thomas D. Seeley.
p. cm.
Includes bibliographical references and index.
ISBN 978-0-691-14721-5 (alk. paper)
1. Honeybee—Behavior. 2. insect societies. I. Title.
QL568.A6S439 2010
595.79'9156—dc22 2010010265

British Library Cataloging-in-Publication Data is available
This book has been composed in Perpetua
Printed on acid-free paper. ∞
Printed in the United States of America
5 7 9 10 8 6

Contents

PROLOGUE

Beekeepers have long observed, and lamented, the tendency of their hives to swarm in the late spring and early summer. When this happens, the majority of a colony's members—a crowd of some ten thousand worker bees—flies off with the old queen to produce a daughter colony, while the rest stays at home and rears a new queen to perpetuate the parental colony. The migrating bees settle on a tree branch in a beardlike cluster and then hang there together for several hours or a few days. During this time, these homeless insects will do something truly amazing; they will hold a democratic debate to choose their new home.

This book is about how honeybees conduct this democratic decision-making process. We will examine the way that several hundred of a swarm's oldest bees spring into action as nest-site scouts and begin exploring the countryside for dark crevices. We will see how these house hunters evaluate the potential dwelling places they find; advertise their discoveries to their fellow scouts with lively dances; debate vigorously to choose the best nest site, then rouse the entire swarm to take off; and finally pilot the cloud of airborne bees to its home. This is typically a hollow tree several miles away.

My motive for writing this book about democracy in honeybee swarms is twofold. First, I want to present to biologists and social scientists a coherent summary of the research on this topic that has been conducted over the last 60 years, starting with the work of Martin Lindauer in Germany. Until now, the information on this subject has remained scattered among dozens of papers published in numerous scientific journals, which makes it hard to see how each discovery is connected to all the others. The story of how honeybees make a democratic decision based on a face-to-face, consensus-seeking assembly is certainly important to behavioral biologists interested in how social animals make group decisions.

I hope it will also prove important to neuroscientists studying the neural basis of decision making, for there are intriguing similarities between honeybee swarms and primate brains in the ways that they process information to make decisions. Furthermore, I hope the story of the house-hunting bees will be helpful to social scientists in their search for ways to raise the reliability of decision making by human groups. One important lesson that we can glean from the bees in this regard is that even in a group composed of friendly individuals with common interests, conflict can be a useful element in a decision-making process. That is, it often pays a group to *argue* things carefully through to find the best solution to a tough problem.

My second motive for writing this book is to share with beekeepers and general readers the pleasures I have experienced in investigating swarms of honeybees. I can thank these beautiful little creatures for many hours of the purest joy of discovery, interspersed among (to be sure) days and weeks of fruitless and sometimes discouraging work. To give a sense of the excitement and challenge of studying the bees, I will report numerous personal events, speculations, and thoughts about conducting scientific studies.

The work described here rests on a solid foundation of knowledge that the late Professor Martin Lindauer (1918–2008) created with his studies of the house-hunting bees in the 1950s. I wish to dedicate this book to Martin Lindauer, my friend and teacher, whose pioneering investigations inspired my own explorations of the wonderland of the bees' society.

Tom Seeley
Ithaca, New York

1

INTRODUCTION

Go to the bee,
thou poet:
consider her ways
and be wise.
—George Bernard Shaw, Man and Superman*, 1903*

Honeybees are sweetness and light—producers of honey and beeswax—so it is no great wonder that humans have prized these small creatures since ancient times. Even today, when rich sweets and bright lights are commonplace, we humans continue to treasure these hard-working insects, especially the 200 billion or so that live in partnership with commercial beekeepers and perform on our behalf a critical agricultural mission: go forth and pollinate. In North America, the managed honeybees are the primary pollinators for some 50 fruit and vegetable crops, which together form the most nutritious portion of our daily diet. But honeybees also provide us another great gift, one that feeds our brains rather than our bellies, for inside each teeming beehive is an exemplar of a community whose members succeed in working together to achieve shared goals. We will see that these little six-legged beauties have something to teach us about building smoothly functioning groups, especially ones capable of exploiting fully the power of democratic decision making.

Our lessons will come from just one species of honeybee, *Apis mellifera*, the best-known insect on the planet. Originally native to western Asia, the Mid-

Fig. 1.1 A comb built of beeswax sculpted into hexagonal cells and filled with pollen from various species of plants.

dle East, Africa, and Europe, it is now found in temperate and tropical regions throughout the world thanks to the dispersal efforts of its human admirers. It is a bee that is beautifully social. We can see this beauty in their nests of golden combs, those exquisite arrays of hexagonal cells sculpted of thinnest beeswax (fig. 1.1). We can see it further in their harmonious societies, wherein tens of thousands of worker bees, through enlightened self-interest, cooperate to serve a colony's common good. And in this book, we will see the social beauty of honeybees vividly, and in fine detail, by examining how a colony achieves near-perfect accuracy when it selects its home.

Choosing the right dwelling place is a life-or-death matter for a honeybee colony. If a colony chooses poorly, and so occupies a nest cavity that is too small to hold the honey stores it needs to survive winter, or that provides it with poor protection from cold winds and hungry marauders, then it will die. Given the vital importance of choosing a suitably roomy and snug homesite, it is not surprising that a colony's choice of its living quarters is made not by a few bees acting alone but by several hundred bees acting collectively. This book is about how this sizable search committee almost always makes a good choice. We will uncover the means by which these house-hunting bees scour the neighborhood for potential nest sites, report the news of their discoveries, conduct a frank debate about these options, and ultimately reach an agreement about which site will be their colony's new dwelling place. In short, we will examine the ingenious workings of honeybee democracy.

There is one common misunderstanding about the inner operations of a honeybee colony that I must dispel at the outset, namely that a colony is governed by a benevolent dictator, Her Majesty the Queen. The belief that a colony's coherence derives from an omniscient queen (or king) telling the workers what to do is centuries old, tracing back to Aristotle and persisting until modern times. But it is false. What is true is that a colony's queen lies at the heart of the whole operation, for a honeybee colony is an immense family consisting of the mother queen and her thousands of progeny. It is also true that the many thousands of attentive daughters (the workers) of the mother queen are, ultimately, all striving to promote her survival and reproduction. Nevertheless, a colony's queen is not the Royal Decider. Rather, she is the Royal Ovipositer. Each summer day, she monotonously lays the 1,500 or so eggs needed to maintain her colony's workforce. She is oblivious of her colony's ever-changing labor needs—for example, more comb builders here, fewer pollen foragers there—to which the colony's staff of worker bees steadily adapts itself. The only known dominion exercised by the queen is the suppression of rearing additional queens. She accomplishes this with a glandular secretion, called "queen substance," that workers contacting her pick up on their antennae and distribute to all corners of the hive. In this way, these workers spread the word that their mother queen is alive and well, hence there is no need to rear a new queen. So the mother queen is not the workers' boss. Indeed, there is no all-knowing central planner supervising the thousands and thousands of worker bees in a colony. The work of a hive is instead governed collectively by the workers themselves, each one an alert individual making tours of inspection looking for things to do and acting on her own to serve the community. Living close together, connected by the network of their shared environment and a repertoire of signals for informing one another of urgent labor needs—for example, dances that direct foragers to flowers brimming with sweet nectar—the workers achieve an enviable harmony of labor without supervision.

Collective Intelligence

This book focuses on what I believe is the most wondrous example of how the multitude of bees in a hive, much like the multitude of cells in a body, work together without an overseer to create a functional unit whose abilities far transcend those of its constituents. Specifically, we will examine how a swarm of honeybees achieves a form of collective intelligence in the choice of its home. As will be described in chapter 2, the bees' process of house hunting unfolds in late spring and early summer, when colonies become overcrowded in their nesting cavities (bee hives and tree hollows) and then cast a swarm. When this happens, about a third of the worker bees stay at home and rear a new queen, thereby perpetuating the mother colony, while the other two-thirds of the workforce—a group of some ten thousand—rushes off with the old queen to create a daughter colony. The migrants travel only 30 meters (about 100 feet) or so before coalescing into a beardlike cluster, where they literally hang out together for several hours or a few days (fig. 1.2). Once bivouacked, the swarm will field several hundred house hunters to explore some 70 square kilometers (30 square miles) of the surrounding landscape for potential homesites, locate a dozen or more possibilities, evaluate each one with respect to the multiple criteria that define a bee's dream home, and democratically select a favorite for their new domicile. The bees' collective judgment almost always favors the site that best fulfills their need for sufficiently spacious and highly protective accommodations. Then, shortly after completing their selection process, the swarm bees implement their choice by taking flight en masse and flying straight to their new home, usually a snug cavity in a tree a few miles away.

The enchanting story of house hunting by honeybees presents us with two intriguing mysteries. First, how can a bunch of tiny-brained bees, hanging from a tree branch, make such a complex decision and make it well? The solution to this first mystery will be revealed in chapters 3, 4, 5, and 6. Second, how can a swirl-

Fig. 1.2 A swarm of honeybees, with approximately ten thousand worker bees and one queen bee.

ing ensemble of ten thousand airborne bees steer themselves and stay together throughout the cross-country flight to their chosen home, a journey whose destination is typically a small knothole in an inconspicuous tree in a remote forest corner? The solution to this second mystery will be revealed in chapters 7 and 8.

We will see that the 1.5 kilograms (3 pounds) of bees in a honeybee swarm, just like the 1.5 kilograms (3 pounds) of neurons in a human brain, achieve their collective wisdom by organizing themselves in such a way that even though each individual has limited information and limited intelligence, the group as a whole makes first-rate collective decisions. This comparison between swarms and brains might seem superficial, but there is real substance here. Over the last two decades, while other sociobiologists and I have been analyzing the behavioral mechanisms of decision making by insect societies, neurobiologists have been investigating the neuronal basis of decision making by primate brains. It turns out there are intriguing similarities in the pictures that have emerged from these two independent lines of study. For example, the studies of individual neuron activity associated with the eye-movement decisions in monkey brains and the studies of individual bee activity associated with nest-site decisions in honeybee swarms have both found that the decision-making process is essentially a competition between alternatives to accumulate support (e.g., neuron firings and bee visits), and the alternative that is chosen is the one whose accumulation of support first surpasses a critical threshold. Consistencies like these suggest that there are general principles of organization for building groups far smarter than the smartest individuals in them. We will explore these principles in chapter 9, where we will compare the decision-making mechanisms of bee swarms and primate brains, and in chapter 10, where we will review the lessons that have been learned from the bees about how to structure a group so that it functions as a smart decision maker.

Group decisions by humans are widespread and important, whether they are small-scale (e.g., agreements made among friends and colleagues), medium-scale (e.g., choices made in democratic town meetings), or large-scale decisions (e.g., national elections or international agreements). Not surprisingly, humans have puzzled over how to optimize group decision making for millennia, at least since Plato's *The Republic* (360 BC) and no doubt long before, and yet many questions remain open about how humans can improve social choice. In chapter 10, I will offer some suggestions, what I call "Swarm Smarts" because they have been learned from the bees, on how human groups can organize themselves to improve their decision making. The American essayist Henry David Thoreau expressed skepticism about the wisdom of crowds when he wrote, "The mass never comes up to the standard of its best member, but on the contrary degrades itself to a level with the lowest." The German philosopher Friedrich Nietzsche was even more negative about group intelligence when he wrote, "Madness is rare in individuals—but in groups . . . it is the rule." Certainly there are many examples of groups making lousy decisions—think of stock market bubbles or of deadly stampedes from burning buildings—but the reality of honeybee swarms making good decisions shows us that there really are ways to endow a group with a high collective IQ.

Dancing Bees

The scientific story told in this book started in Germany almost seventy years ago, in the summer of 1944, when a distinguished professor of zoology at the University of Munich, Karl von Frisch, made a revolutionary discovery for which he would eventually receive the Nobel Prize: an insect, the worker honeybee, can inform her hive mates of the direction and distance to a rich food source by means of dance behavior. Von Frisch had already known for nearly thirty years that when a lone forager finds a rich source of nectar, she returns excitedly to her hive and performs a conspicuous "waggle dance." In performing this eye-catching behavior, the dancer walks straight ahead on the vertical surface of a comb, waggling her body from side to side, then she stops the "waggle run" and turns left or right to make a semicircular "return run" back to her starting point, whereupon she produces another waggle run followed by another return run, and so on (fig. 1.3). Each waggle dance consists, therefore, of a series of dance circuits, and each dance circuit contains a waggle run and a return run. Von Frisch also knew that a bee may continue dancing for some seconds or even some minutes, all the while trailed by unemployed foragers that, in his own words, "take part in each of her manoeuvrings so that the dancer herself, in her madly wheeling movements, appears to carry behind her a perpetual comet's tail of bees." Furthermore, he knew well that after a dance-follower has tripped along behind a dancer throughout several circuits of her dance, she rushes out of the hive to search for the bonanza announced by the dancing bee. But before 1944, von Frisch thought that the only thing the dance-followers learned from the dancer was the fragrance of the flowers she had visited—which they detected by holding their antennae close to the dancer to smell the floral scents adhering to her body—and that upon leaving the hive the newly aroused bees simply searched in ever-expanding circles until they discovered flowers with the memorized fragrance. What von Frisch discovered in 1944 was nearly incredible: the dance-followers did not search for flowers with the matching scent everywhere around the hive, but only in the vicinity of where the dancer had foraged, even if she had foraged in a remote spot, such as along a shady lakeside trail far from the hive. Without a doubt, the newcomers were somehow acquiring from the successful forager information about food-source

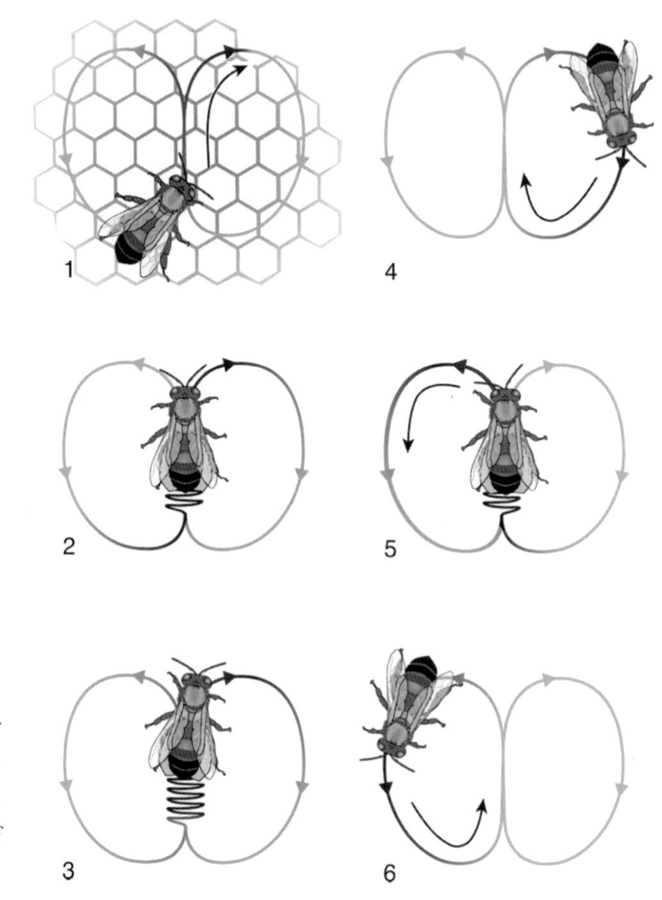

Fig. 1.3 The movement pattern of a worker bee performing a waggle dance on the vertical surface of a comb inside her colony's hive. The bee is shown performing two circuits of the waggle dance.

location as well as food-source scent. Could this location information be communicated inside the hive, by means of the bees' dances?

The answer turned out to be a definitive *yes*. In the summer of 1945, amid the chaos in Europe following the end of World War II, von Frisch returned to his dancing bees, now observing their movements more closely than ever before, examining them for clues that would help him solve his mystery. He discovered that when a bee performs a waggle run inside a dark hive, she produces a miniaturized reenactment of her recent flight outside the hive over sunlit countryside, and in this way indicates the location of the rich food source she has just visited (fig. 1.4). Her encoding of the information about food-source location works as

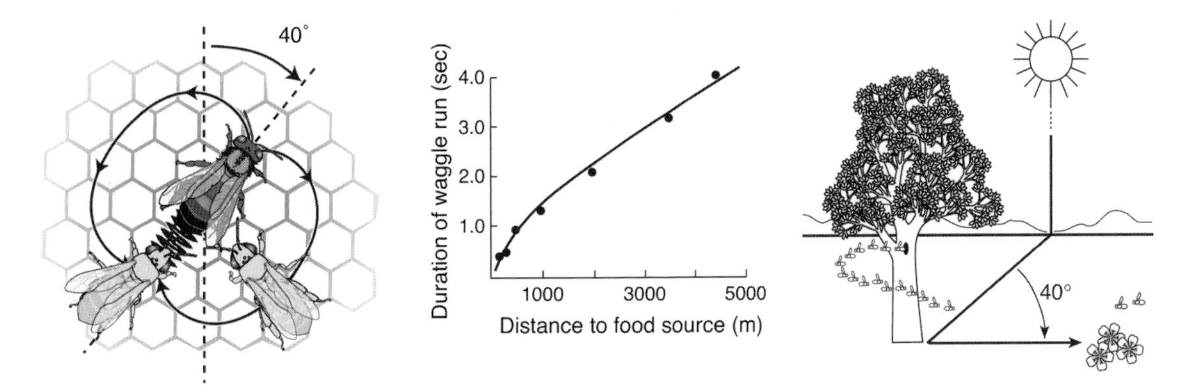

Fig. 1.4 How a dancing bee encodes information about the distance and direction to a rich patch of flowers. Distance coding: The duration of each waggle run is proportional to the length of the outbound flight. Direction coding: Outside the hive the bee notes the angle of her outbound flight relative to the sun's direction, and then inside the hive she orients her waggle runs at the same angle relative to straight up on the comb. Two followers are acquiring the dancing bee's information.

follows. The duration of the waggle run—made conspicuous despite the darkness by the dancer audibly buzzing her wings while waggling her body—is directly proportional to the length of the outward journey. On average, one second of the combined body-waggling/wing-buzzing represents some 1,000 meters (six-tenths of a mile) of flight. And the angle of the waggle run, relative to straight up on the vertical comb, represents the angle of the outward journey relative to the direction of the sun. Thus, for example, if a successful forager walks directly upward while producing a waggle run, she indicates that "the feeding place is in the same direction as the sun." Or, if the waggling bee heads 40 degrees to the right of vertical, her message is, "The feeding place is 40 degrees to the right of the sun," as shown in figure 1.4. Perhaps most remarkably, the bees that follow a dancer, monitoring her waggle runs, are able to decode her dance and put her flight instructions into action.

While von Frisch was deciphering the secret message of the waggle dance, he was also supervising a young graduate student, named Martin Lindauer, who was to prove von Frisch's most gifted disciple in revealing the inner workings of

Fig. 1.5 Karl von Frisch (elderly gentleman, center), Martin Lindauer (young man, leftmost), and other students as they prepare to perform an experiment with the bees, circa 1952.

honeybee colonies (fig. 1.5). Lindauer is an especially important figure in this book, for he pioneered the study of honeybee democracy as practiced in a bee swarm choosing its home.

Lindauer was born in a tiny village in the foothills of the Bavarian Alps, the next to youngest of 15 children in a poor farming family. Here he grew up close to nature—including the bees in his father's hives—but he was also an outstanding student and won a scholarship to a distinguished boarding school in Landshut, Germany. In April 1939, eight days after his high school graduation, he was drafted into Hitler's Work Service to dig trenches, and six months later he was transferred to the army and assigned to an antitank unit. In July 1942, during fierce fighting on the Russian front, he received deep shrapnel wounds from an exploding grenade. This proved his salvation. He was removed from the front, but the other 156 men in his company went on to the battle of Stalingrad. Only three returned alive.

While recovering in Munich, Lindauer was recommended by his physician to visit the university and attend one of the General Zoology lectures delivered by the famous Professor Karl von Frisch. Lindauer later recalled that when he did, and saw von Frisch talking about cell division, he returned to "a new world of humanity," where people create rather than destroy. He resolved to study biology, and in the summer of 1943, after being discharged from the army as a severely wounded soldier, he began university studies in Munich. Ultimately, he started his PhD research on honeybees, with von Frisch his advisor, in the spring of 1945.

Dirty Dancers

Lindauer had a knack for noticing little things in passing—some curious anomaly or surprising behavioral quirk—that would eventually turn out to be important. It was thanks to this special talent that Lindauer embarked on his studies of house hunting by honeybees, which he would later refer to as "the most beautiful experience" in his scientific work. It all began on a spring afternoon in 1949, when Lindauer was walking past the beehives outside the Zoological Institute, and he spied a golden mass of bees, a swarm, hanging on a bush. Pausing to study it, he noticed several bees performing waggle dances on the swarm's surface, doing so with their usual attention-grabbing vigor, but instead of striding across a beeswax comb, the normal dance floor for bees, these dancers were walking over the backs of other bees. Initially, he assumed that these swarm dancers were foragers bringing food back to the swarm, because all the dancing bees that he and von Frisch had studied over the past few years had been foragers bringing food back to the hive. But with his customary patience in watching the bees, Lindauer lingered by the swarm, kept observing its dancers, and gradually realized that they did not look like foragers, for unlike pollen foragers they never carried pollen loads and unlike nectar foragers they never regurgitated droplets of nectar to adjacent bees. He also saw something odd: many of the dancers arrived at the swarm dirty and dusty. When he plucked several of these rather grubby bees from the swarm with forceps, dusted them off with a small paintbrush, and examined the debris under a microscope, he found no pollen grains, just various forms of dirt particles. "Black with soot," he reported, "red with brick dust, white with flour, or gray and dusty as if they had been grubbing a hole in the ground." Sniffing the bees black with soot, he was reminded of the aroma of a chimney sweep.

Lindauer concluded that these dusty and dirty dancers were certainly not foragers. He suspected that they were nest-site scouts who had discovered potential nesting cavities amid the rubble of bombed-out Munich—an unused chimney here, a cavity in some collapsed brick wall there, or even a forgotten flour chest in an abandoned attic—and were indicating the locations of their discoveries by performing waggle dances. He was eager to test this hunch by making further observations of swarm bees, but in 1949, with Germany's economy still in shambles and von Frisch's laboratory short of bees, von Frisch directed the institute's beekeeper to promptly hive all swarms, lest the bees be lost. This meant that the bees' house-hunting process was cut short. And so, for the time being, Lindauer's study of the house-hunting process was also cut short. But he persisted in seeking approval to leave some swarms alone so he could study their dancing bees, and two summers later, in 1951, von Frisch granted Lindauer permission to study as he wished all the swarms from the beehives kept in the garden of the Zoological Institute in Munich.

Later, in chapters 3 to 6, we will review in detail the intriguing story of democratic decision making by honeybees that Lindauer started piecing together in 1951. For now, we will consider only how he tested his hypothesis that the bees performing dances on a swarm are nest-site scouts advertising potential homes. In the summer of 1951, he examined the dances on nine swarms. Sitting patiently beside each swarm for hours and days on end, he labeled each dancing bee with a paint dot when she began her dancing, and he noted the direction and distance of the site indicated by her first dance. (Lindauer reasonably assumed that his swarming bees were encoding distance and direction information in their dances in exactly the same way as von Frisch had found for his foraging bees.) These swarm-side vigils led Lindauer to a surprising finding: when dancers started to appear on a swarm, they announced a dozen or more widely separated locations, but after a few hours or a few days, they began to announce one location in in-

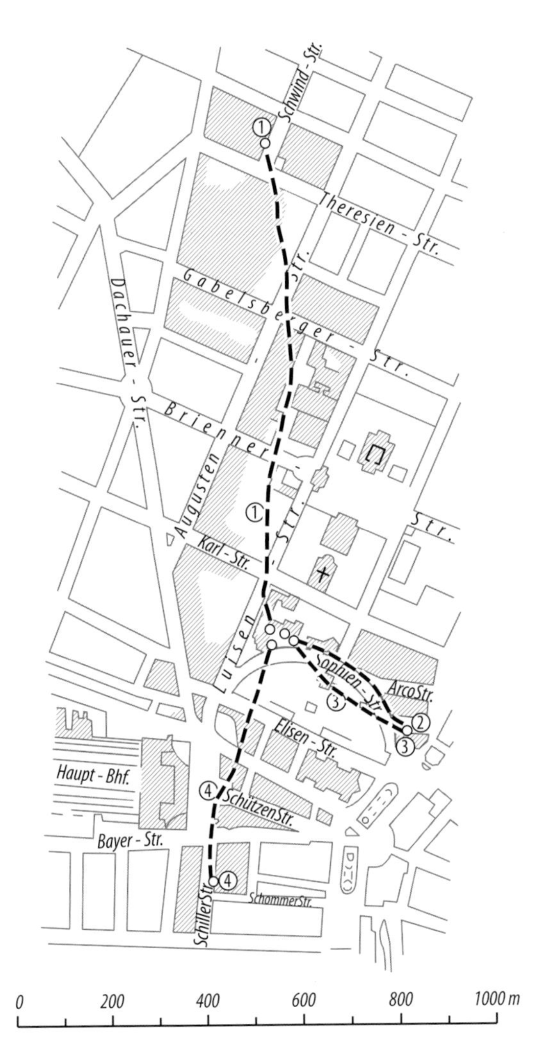

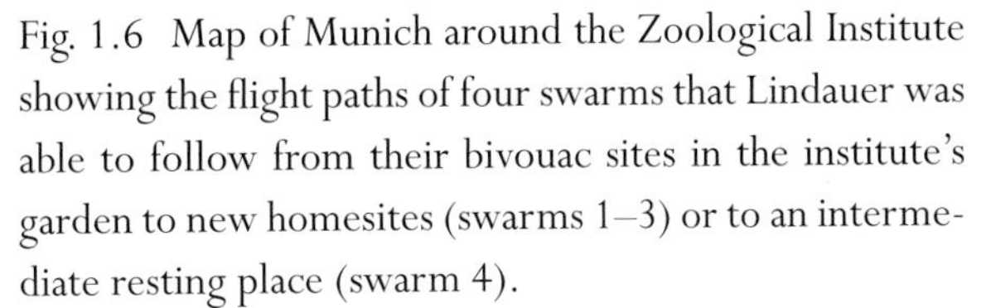

Fig. 1.6 Map of Munich around the Zoological Institute showing the flight paths of four swarms that Lindauer was able to follow from their bivouac sites in the institute's garden to new homesites (swarms 1–3) or to an intermediate resting place (swarm 4).

creasing numbers. Ultimately, during the last hour or so before the swarm took off to fly to its new home, the bees dancing on the swarm all indicated just one distance and direction. Lindauer reasoned that if the bees dancing on a swarm had been searching for nest sites, and if they were performing dances to advertise their finds, then the location they were indicating unanimously in the end should correspond to the location of the swarm's new residence. To test this prediction, he tried to follow each swarm when it flew to its new home, by sprinting down Munich's streets and alleys beneath the airborne swarm (fig. 1.6). Three times he succeeded! And in each case the spot indicated by the bees' closing dances matched the address of their new dwelling place. So, sure enough, Lindauer's dirty dancers really were house hunters.

Catching the Buzz

In June 1952, when Lindauer was busy in his second summer of swarm watching in Munich, I was born some 6,500 kilometers (4,000 miles) away, in a little town in Pennsylvania. A few years later, my family moved to Ithaca, New York, which has been basically my hometown ever since. While growing up in Ellis Hollow, a rural community a few miles east of Ithaca, I spent much time alone exploring the wild areas around our home: shady hardwood forests on the steep hillsides, sunlit abandoned fields where the land slopes gently, and winding Cascadilla Creek connecting broad swamps in the valley bottom. My favorite find was about a mile from home down a dirt road that led to an old farmhouse. Here, in a sunny spot beside a field of goldenrod, I discovered two wooden hives of bees that belonged to a beekeeper. I loved visiting these hives. Sitting beside one, I could see bees landing heavily at the entrance with loads of brightly colored pollen, I could hear the hum of bees fanning their wings to ventilate their nest, and I could smell the aroma of ripening honey. That thousands of insects could live together so densely and harmoniously, and could build delicate wax combs filled with delicious honey, was an almost miraculous wonder that left a deep impression. No less impressive was what I saw when I lay in the tall grass beside these hives: thousands of humming bees crisscrossing the blue summer sky like shooting stars.

It was not until high school—when my peers seemed interested mostly in sports, motorcycles, and girls—that I became utterly spellbound by the bees. I'd had a deep curiosity about honeybees ever since a beekeeper made a show-and-tell visit to my third grade class in elementary school, and I had especially enjoyed earning an Insect Study merit badge as a Boy Scout in junior high school. I had even daydreamed from time to time about ordering a hive and some bees from the Sears catalog and taking up beekeeping. But I didn't really catch the buzz until the summer day in 1969 when I found a swarm hanging from a tree branch, quickly nailed together some boards to make a crude hive, shook the bees into my hive, and brought them home. Now I had these little sparks of wonderment living in a box that I could gently open to watch them closely, and I did so for several hours each day after work, mesmerized by the intricate behaviors of the individual bees and by the peace of their great community.

In the fall of 1970, when I began my freshman year at Dartmouth College, I hadn't yet realized that studying bees could be serious business, and I aimed to become a physician who would do beekeeping as a hobby. But the allure of the bees grew stronger and stronger. I made bees or beekeeping the subject of nearly every college writing assignment. I chose chemistry as my major area of study so that I could become a cryptographer of the chemical (pheromonal) language of bees, which was just starting to be deciphered. And I returned each summer to Ithaca so that I could work at the Dyce Laboratory for Honey Bee Studies at Cornell. The director of this lab, Roger A. Morse ("Doc"), understood my passion for the bees and advised me to give graduate school some thought. During my last two years at Dartmouth, I gradually became aware that my interests in entomology had eclipsed those in medicine, so even though I applied to and was accepted by three medical schools, I was thrilled when I was accepted at Harvard to pursue graduate studies with the noted insect sociologist Edward O. Wilson, whose 1971 book *The Insect Societies* had made an enormous impression on me.

When I arrived at Harvard in the fall of 1974, I had the good fortune of being assigned to Bert Hölldobler, a brilliant investigator of ant behavior and a young man of immense personal appeal, for my provisional thesis advisor. Bert had recently moved from the University of Frankfurt, Germany. He had been hired by Harvard as a full professor to seed the university with the Karl von Frisch approach to animal behavior study: close behavioral observations of animals living in nature combined with incisive experimental investigations on the mechanisms underlying their behavior. In Germany, Bert had studied under Martin Lindauer, so he knew the bees as well as the ants, his own first love. He supported my beeophilia, and we were quick to become friends.

Bert Hölldobler's connection with Martin Lindauer was important to me because I knew that I wanted my PhD thesis research to deepen Lindauer's investigations of how a honeybee colony works as unit, a kind of superorganism. I

Fig. 1.7 The author in 1974, making a pilot study of a swarm choosing its home.

was especially keen to analyze in greater detail the decision-making process of a honeybee swarm. Back at Dartmouth, I had read Lindauer's little book, *Communication among Social Bees*, and had been especially intrigued by chapter 2, titled "Communication by Dancing in Swarm Bees," in which he summarized his studies of how a swarm chooses its home. Indeed, I was so fascinated that I tracked down Lindauer's full report on this work, a 62-page paper, all in German, titled "Schwarmbienen auf Wohnungssuche" (Swarm bees out house hunting). There was just one problem, I could not read German. My solution was to enroll in an introductory German course at Dartmouth, buy a German-English dictionary, make a Xerox copy of Lindauer's paper, and patiently decipher Lindauer's big paper. (I penciled in the paper's margins the meaning of every new German word and this densely annotated copy, now 38 years old, is a prize part of my reprint collection.) In poring over this paper, I began to realize that Lindauer's pioneering investigation of the swarm bees' process of collective decision making provided just a preliminary understanding of the subject, and that it, like all good scientific studies, raised many more questions than it answered. I was also amazed—and I confess delighted—that during the nearly 20 years since Lindauer had published his work in 1955, no one had pressed the investigation more deeply. I resolved to do so, starting with my PhD thesis research (fig. 1.7).

This book aims to present to biologists and general readers what was learned by Lindauer in the 1950s, and by myself and others since the 1970s, about how the workers in a honeybee swarm conduct a democratic decision-making process to make the life-or-death choice of where to build their new home. This work has revealed how the process of evolution by natural selection, operating over millions of years, has shaped the behavior of bees so that they coalesce into a single collective intelligence. This story about the bees also provides useful guidelines to human groups whose members share common interests and want to make good group decisions. Mainly, though, this book tries to be a window into the private world of a honeybee colony. If it bolsters, in any way, an appreciation of these little creatures for the beauty of their social behavior, along with their service in keeping the world flowering and fruitful, then it will have achieved its purpose.

 2 

LIFE IN A HONEYBEE COLONY

. . . this being an Amazonian
or feminine kingdome.
—Charles Butler, The Feminine Monarchie, *1609*

The honeybee, *Apis mellifera*, is just one of nearly 20,000 species of bees that exist worldwide. They are a surprisingly diverse lot—some are smaller than a rice grain, while others will half fill a teacup—but they are all descended from one ancestral species of vegetarian wasp that lived approximately 100 million years ago, in the Early Cretaceous period, when huge dinosaurs were still stomping about and flowering plants were just starting to appear. Even today, many bee species are remarkably wasplike in appearance, but behaviorally the two groups are distinct. Nearly all wasps, including the familiar paper wasps and yellow jackets, are predators that kill other insects or spiders (often by stinging) to provide the egg-laying females and their growing young with proteinaceous food. Bees, however, have abandoned the carnivorous behavior of their ancestors and depend instead on collecting protein-rich pollen from flowers. This pollenivorous habit explains the decidedly fuzzy, almost teddy-bearish, look of many bees; their bodies are thickly covered with plumose hairs that efficiently catch pollen grains when a bee scrabbles through a flower.

Both bees and wasps regularly visit flowers, for both types of insects feed on sugary nectar for energy, but it is between the pollen-loving bees and the flowering plants that a strong mutual dependence has evolved over the millions of years since both groups arose. These days, the two are made for each other. Bees depend on flowers for adequate nourishment, while many flowering plants depend on bees for sexual reproduction. Bees, with their hairy bodies and fixation on flowers as protein sources, serve as flying penises for the plants, picking up pollen grains from the bursting anthers of one flower and depositing them on the sticky stigma of another. Introducing a hive of honeybees to any flowering area—garden, orchard, blooming wayside, or prairie—brings to the neighborhood, in effect, a large, dawn-to-dusk "escort" service of the flowers' little friends.

Honeybees are unusual bees in that they live in teeming societies whose massive nests of honeycomb fill the boxy hives of beekeepers or, as we shall see, suitably spacious cavities in hollow trees. In most bee species, by contrast, individuals live in solitude and build small nests in narrow tunnels excavated inside plant stems or in sandy soil. The typical life history of one of these solitary bee species starts in late spring or early summer when a mated female emerges from an overwintering burrow (the males having died off the previous fall). Over the next few weeks, this motherly bee will excavate a multichambered nest, provision each chamber with a sticky ball of pollen moistened with nectar, lay one whitish egg atop each pollen ball, seal up each chamber, and then leave her offspring to eat their way to adulthood later that summer. She will die long before her offspring emerge as adults, mate with one another, and prepare for the coming winter. Clearly, most bees are loners.

A Composite Being

When we look through the glass walls of a honeybee observation hive, or gently lift the lid from a conventional beehive and peer inside, we see the opposite of loner bees: thousands and thousands of bees living together. Virtually all are female worker bees, all of whom are daughters of the one queen bee that lives in their midst. Even though these workers are females and are fully equipped to care for offspring, they have poorly developed ovaries and they rarely lay any eggs. If we proceed to search the combs of the hive carefully, we will eventually locate the queen, who resembles the workers but is a bit bigger, with a longer abdomen and longer legs (fig. 2.1). Her greater size is impressive, but what most

Fig. 2.1 The queen is larger than the surrounding workers, who feed and groom her.

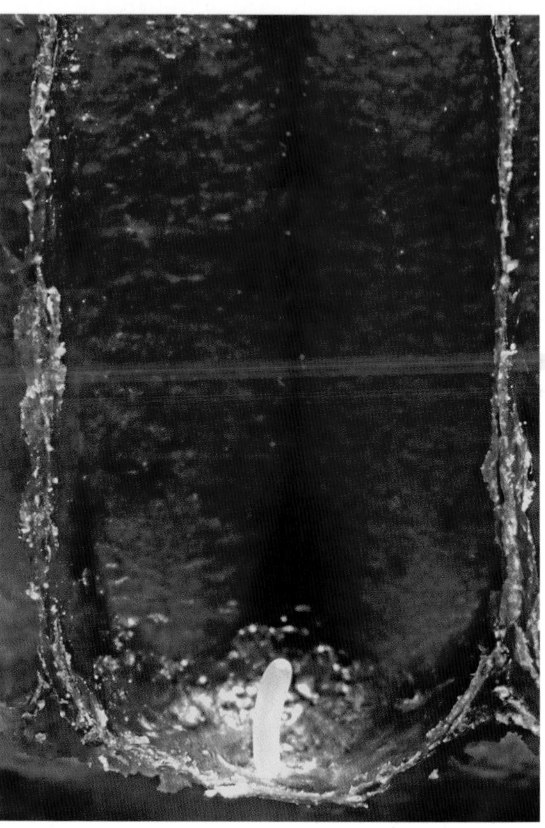

Fig. 2.2 When the queen finds a clean, empty cell, she inserts her long abdomen and lays a single egg at the base of the cell.

renders her conspicuous is how she moves slowly, indeed majestically, across the combs and how she is treated by her worker daughters. As the queen advances, the workers before her step back to clear her path, and when she pauses, the dozen or so workers beside her gingerly step forward to feed and groom her, forming a retinue of nudging bees that encircles her completely. In contrast to the workers, the queen is an amazing layer of eggs, depositing them in cells at a rate of one or more per minute, or more than 1,500 per day (with a combined weight nearly equal to her own) in late spring and early summer, when a colony's brood rearing is at its peak (fig. 2.2). Over an entire summer, a colony's queen will produce some 150,000 eggs, hence about half a million during the two or three years of her likely lifespan.

Most of the pearly white eggs the queen lays will be fertilized, but some will be unfertilized. During the first week of her life, she flew from her colony's hive and mated with 10 to 20 males from other hives in the area, and so procured a lifetime supply of approximately five million sperm. The queen stores all these sperm in suspended animation inside a spherical organ called a spermatheca, which lies in the rear of her abdomen, behind her massive ovaries. With each egg she lays, the queen decides whether to dispense a few fertilizing sperm or to hold them back, and in this way she determines the sex of her offspring: fertilized for female, unfertilized for male. Whether a fertilized egg develops into a nonbreeding worker or an egg-laying queen depends on how it is treated. If it is deposited in a standard-size cell in the combs, where after hatching into a larva it will be fed by the workers with standard-quality larval food, then it will develop into a worker. But if a fertilized egg is deposited in a large, specially built queen cell, hanging from the bottom of a comb, then the larva it gives rise to will be fed a lavish diet of nutrient-rich secretions (so-called royal jelly), and its development will be channeled to a developmental pathway that produces a queen. For the fertilized eggs of bees, food is destiny.

A queen withholds sperm from less than 5 percent of her eggs, but these unfertilized eggs are important for they give rise to her sons, the colony's drones (fig. 2.3). These are a colony's brawniest bees, endowed with huge eyes for spying young queens out on nuptial flights and massive flight muscles for chasing after the queens at speeds up to 35 kilometers an hour (about 22 miles per hour).

Fig. 2.3 The three types of adult honeybees.

They are also a colony's laziest bees. Unlike the workers who perform all the household tasks inside their hive—clean cells, feed larvae, build combs, ripen honey, ventilate hive, guard entrance, etc.—the drones spend their time at home simply hanging out in restful leisure, from time to time helping themselves to meals from the colony's honey reserves or begging feedings from their worker-sisters. Nevertheless, they make a fundamentally important contribution to their colony's success, for in mating with the young queens of the neighboring colonies they help their colony win in the ceaseless evolutionary competition to pass genes on to future generations. And when it comes to seeking sex, drone honeybees are no slackers. Every sunny afternoon, once a drone reaches sexual maturity at about 12 days of age, he will fly from his hive looking for action. In ways that remain mysterious, he will find his way to one of the traditional honeybee mating areas ("drone congregation areas") within a few miles of his home, and will fly about this aerial pickup spot, waiting for a young queen to appear. If one does, he will zoom after her. And if he manages to outrace his rivals and contact the queen, he will inseminate her in flight, 10 to 20 meters (30 to 60 feet) up in the sky. If he doesn't make contact, he'll fly home, rest and refuel, and come out later to try his luck again.

One way to think of a honeybee colony is, then, as a society of many thousands of individuals: the queen, workers, and drones just discussed. But to understand the distinctive biology of this species of bee, it is often helpful to think of a colony in a slightly different way, not just as thousands of separate bees but also as a single living entity that functions as a unified whole (fig. 2.4). In other words, it can help to think of a honeybee colony as a superorganism. Just as a human body functions as a single integrated unit even though it is a multitude of cells, the superorganism of a honeybee colony operates as a single coherent whole even though it is a multitude of bees. The fact that both perspectives—colony as superorganism *and* colony as society—are valid reflects the way in which evolution has repeatedly built higher-level units of biological organization: by assembling unified societies of lower-level units. For example, during the origins of multicellular organisms, natural selection favored some societies of cells whose members cooperated rather than competed. Bit by bit, this selection for close cooperation produced the thoroughly integrated societies of cells that we know today, for example, as hummingbirds and human beings. The same sort of selection for extreme cooperation also happened with some societies of animals to produce the thoroughly harmonious, smoothly running insect societies that we can call superorganisms. These include not just colonies of honeybees but also the gigantic colonies of leafcutter ants, driver ants, or fungus-growing termites.

A colony of honeybees is, then, far more than an aggregation of individuals, it is a composite being that functions as an integrated whole. Indeed, one can accurately think of a honeybee colony as a single living entity, weighing as much as 5 kilograms (10 pounds) and performing all of the basic physiological processes that support life: ingesting and digesting food, maintaining nutritional balance, circulating resources, exchanging respiratory gases, regulating water content, controlling body temperature, sensing the environment, deciding how to behave, and achieving locomotion. Consider, for example, the control of body (colony) temperature (fig. 2.5). From late winter to early fall, when the workers are rearing brood, a colony's internal temperature is kept between 34° and 36°C (93° and 96°F)—just below the core body temperature of humans—even as the

Fig. 2.4 A honeybee colony, both a society and a superorganism.

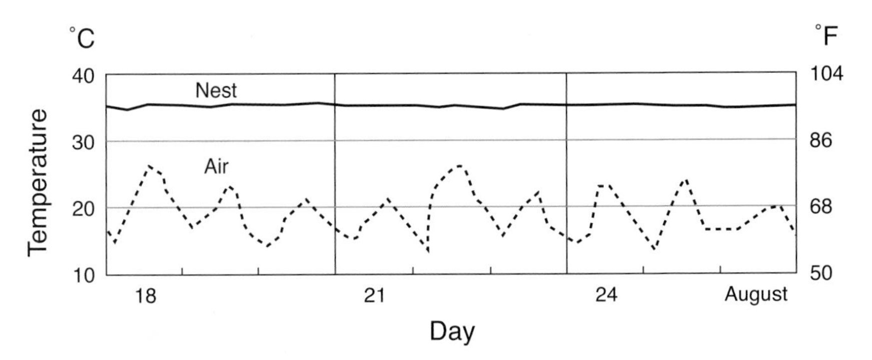

Fig. 2.5 The elevated, and stable, temperature inside a honeybee colony's nest, compared to the outside air temperature.

ambient air temperature ranges from -30° to 50°C (-20° to 120°F). The colony accomplishes this by adjusting the rate at which it sheds the heat generated by its resting metabolism and, in times of extreme cold, by boosting its metabolism to intensify its heat production. A colony's metabolism is fueled by the honey it has stored in its hive. Other indicators of the high functional integration of a honeybee colony include *colonial breathing*: limiting the buildup of the respiratory gas CO_2 inside the hive by increasing its ventilation when the CO_2 level reaches 1–2 percent; *colonial circulation*: keeping the heat-producing bees in the central, brood-nest region of the hive properly fueled with honey carried in from peripheral honey combs; and *colonial fever response*: mounting a disease-fighting elevation of the nest temperature when a colony suffers a dangerous fungal infection of the brood bees. I suggest, though, that the single best demonstration of the superorganismic nature of a honeybee colony is the ability of a honeybee swarm to function as an intelligent decision-making unit when choosing its new home.

Unique Annual Cycle

The key to understanding why honeybee swarms are meticulous in the choice of their living quarters is the unique annual cycle of the honeybee, which depends critically on colonies occupying nesting cavities that are both snug and roomy. Unlike all the other social insect species that live in cold climates, honeybees do

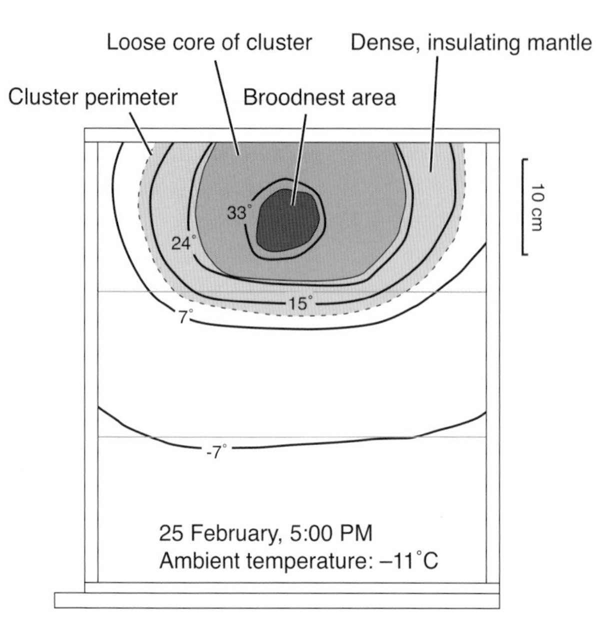

Fig. 2.6 Anatomy of a winter cluster of honeybees.

Fig. 2.7 The nest of a colony that failed to occupy a protective nest cavity.

not survive winter in dormancy, but as fully functioning colonies in self-heated nests. To achieve this means of winter survival, each colony contracts in winter into a tight, well-insulated cluster of bees about the size of a basketball. The cluster's surface temperature is maintained above 10°C (50°F), which is a few degrees above a worker bee's chill-coma threshold, and so is warm enough to keep the outermost bees alive (fig. 2.6). Heat is generated within the cluster by the bees isometrically contracting their two sets of flight muscles (one for elevating the wings and one for depressing them) thereby producing much heat but few or no wing vibrations. These flight muscles endow a bee with a surprisingly powerful means of heat production. Bees fly, of course, by flapping their wings—the most energetically demanding mode of animal locomotion—and the flight muscles of insects are among the most metabolically active of tissues. Indeed, a flying bee expends energy at a rate of about 500 watts per kilogram (250 watts per pound), whereas the maximum power output of an Olympic rowing crew is only about 20 watts per kilogram (10 watts per pound). At any moment, however, only a small portion of the clustered bees will be shivering with maximum intensity, so the total heat output by the approximately two kilograms (four pounds) of bees in a winter cluster isn't 1,000 watts, but is only about 40 watts, a rate of heat production like that of a small incandescent light bulb. In a snug cavity, sheltered from heat-robbing winds, a colony with this level of heat output will survive the winter quite nicely. The importance of inhabiting a protective cavity is demonstrated by the sad fate of the occasional colony that fails to find shelter and nests in the open (fig. 2.7); almost certainly, it will perish when winter's cold arrives.

A honeybee colony runs year-round on flower power, for what fuels a colony's heat production all winter long is the 20 or more kilograms (44+ pounds) of honey that the colony stockpiled in its honeycombs over the previous summer. If one mounts a hive of bees on scales and takes a weight reading each day for an entire year, one will see that winter is a time of steady weight losses as a colony consumes its honey stores, and that summer is a time of episodic weight gains as a colony scrambles to replenish these stores (fig. 2.8). For example, in Ithaca, New York, my colonies restock their honeycombs mainly during the 60-day period between May 15 and July 15, when there unfolds a succession of mass flowerings by plants that produce copious nectar, including black locust and basswood

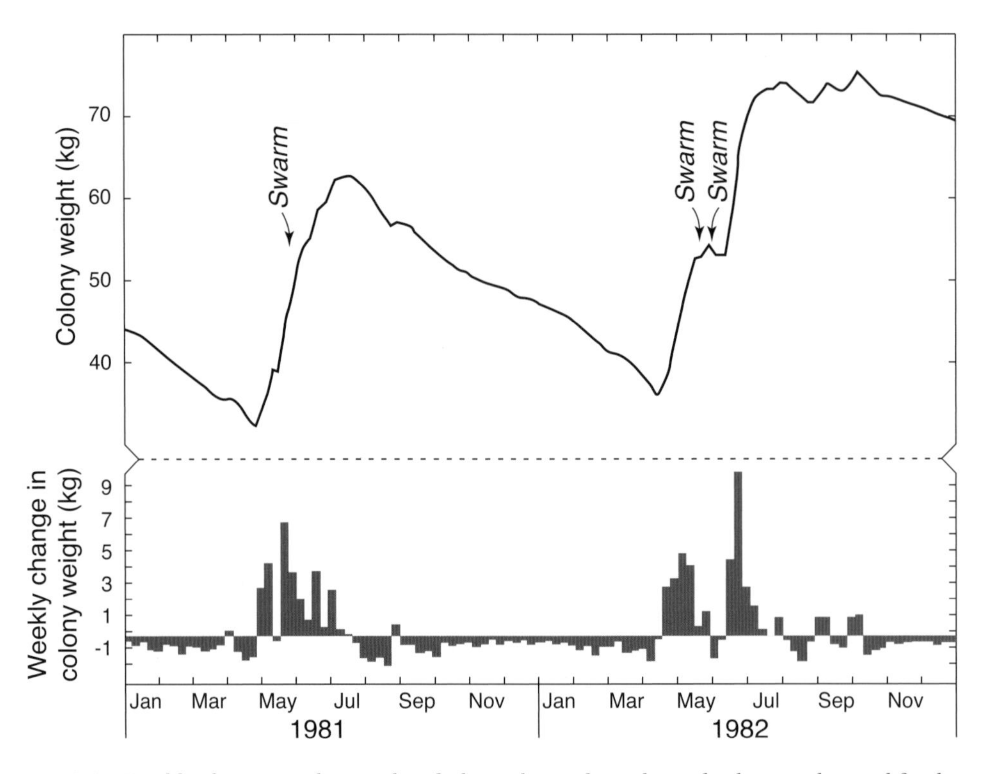

Fig. 2.8. Weekly changes in the weight of a honeybee colony (hive plus bees and stored food).

trees, sumac shrubs, and various herbaceous plants such as dandelions, raspberries, milkweeds, and clover. On a day when the air is warm, the sun is strong, and nectar is flush, the hive that I keep at home on a set of platform scales will grow heavier by several kilograms, virtually all of it fresh honey. Beekeepers call a string of such days a "honey flow."

The task of amassing within a short summer season an ample supply of winter heating fuel is one of the greatest problems faced by a honeybee colony. Honey is a dense, energy-rich food, but even so, 20 kilograms (44 pounds) of the stuff will nearly fill a 16-liter (14-quart) bucket, or more than 50 of those plastic honey bears one sees lined up beside the grape jelly at the supermarket. How much work effort and storage space is needed to create such a bulky hoard of calories? Regarding work effort, given that freshly collected nectar is (on average) a 40 percent sugar solution and fully ripened honey is roughly an 80 percent sugar solution, and given that a foraging bee typically brings home a nectar load weighing about 40 milligrams (0.001 ounces), we can calculate that the collection of enough nectar to produce 20 kilograms (44 pounds) of honey requires more than 1 million foraging trips by a colony's workers. And when one also considers the miles flown and countless blossoms visited on each foraging trip, one realizes what prodigious efforts the bees make over summer to sustain their colony through winter.

Regarding storage space, given that it takes 250 square centimeters of honeycomb to store one kilogram of honey (i.e., 18 square inches of comb per pound of honey), and given that every 250 square centimeters of honeycomb require about 0.9 liters of nest cavity space (to accommodate the honey-filled comb and the adjacent passageways for the bees), we can calculate that the storage of 20 kilograms (44 pounds) of honey requires a nesting cavity of at least some 18 liters (4 gallons). Thus we can see that when a colony chooses its future homesite, it will need to reject tree cavities smaller than these volumes. Ideally, it will find a nesting cavity somewhat roomier still, to accommodate extra honey-filled combs and still more combs for the colony's brood rearing operation, which in spring can fill more than half the cells in a colony's nest as the colony rebuilds its workforce in preparation for swarming. Beekeepers, by the way, have found a clever way to exploit the bees' drive to fill their nests with honey. By housing their colonies in hives that provide vastly more nesting space—about 160 liters (some 36 gallons)—than is needed by bees living in nature, beekeepers induce their colonies to amass astonishing amounts of honey, sometimes more than 100 kilograms (220 pounds) of honey per hive in a summer. Thus a colony of hardworking bees residing in a beekeeper's hive will often provide its landlord with dozens of combs brimming with honey.

The honeybee's annual cycle is unique in other ways besides the overwintering process. Consider how a colony starts rebuilding its workforce in the middle of winter. Shortly after the winter solstice, when the days begin to grow longer but snow still blankets the countryside, each honeybee colony raises the core temperature of its winter cluster to about 35°C (95°F), the optimum temperature for rearing new bees. With the cluster's core now serving as a cozy incubator, the queen begins to lay eggs, using cells that were emptied of their honey during

Fig. 2.9 Cells containing larvae, visible as white, C-shaped grubs.

the preceding weeks of cold. Larvae hatch from these eggs, approximately three days after they were laid, and are fed by the adult workers. At first, the workers feed the larvae a proteinaceous food secreted by glands located in their (the adult bees') heads, but after about three days they wean them to a mixture of honey and pollen. About ten days after hatching from its egg, each larva has grown to nearly fill its cell (fig. 2.9) and starts to spin the cocoon in which it will metamorphose into an adult bee. The adult workers build a wax capping over the cell to protect the immature bee during this delicate, pupal stage of development. Once metamorphosis is complete, in about another week, the fully developed worker bee chews through the capping on her cell and joins her colony's growing workforce. When a colony starts its impressive performance of rearing brood in the middle of winter, there are only a hundred or so cells containing developing bees, but by early spring, when the first flowers blossom, over one thousand cells hold developing bees, and the pace of colony growth quickens daily. Come late spring, when most other insects are just starting to become active, honeybee colonies have already grown to full size, twenty or thirty thousand individuals, and have begun to reproduce.

Colony Reproduction

Reproduction by a honeybee colony is a curiously complex affair, for each colony is a hermaphrodite, meaning that it has both male and female reproductive powers. This is decidedly different from ourselves and most other animals, where each individual is either a male or a female, but it is strikingly similar to many plants, such as apple trees. In fact, to make sense of how a honeybee colony reproduces, I find it helpful to compare how honeybee colonies and apple trees go about achieving sexual reproduction. Their basic similarity is that both types of individual—colony and tree—produce both male and female reproductive propagules. The male propagules are drone bees and pollen grains, whereas the female propagules are queen bees and egg cells. And just as the pollen grains from one apple tree fertilize the egg cells of other trees to create embryos inside seeds that will grow into new trees, the drones from one honeybee colony fertilize the queens of other colonies to create inseminated queens that will give rise to new colonies. Thus both trees and colonies rely on cross-fertilization to avoid the problems associated with inbreeding.

Colonies and trees also resemble each other in how the male and female sides of reproduction differ. For both bees and trees, the male side of reproduction is straightforward. In late spring and early summer, each colony or tree produces a great number of male propagules—thousands of drone bees per colony and millions of pollen grains per tree—that disperse over the countryside and achieve fertilizations. Any one drone bee or pollen grain has a low probability of fertilizing a queen or egg, but because a healthy individual (colony or tree) launches a huge squadron of male propagules, it has a high probability of achieving reproductive success via its small, male gene carriers.

Turning to the female side of reproduction, we find a more complex process in both colony and tree. In each, the fertilized propagule (queen bee or egg cell) is not discharged "naked," as happens with the male propagules, but instead is packaged inside a large and intricate dispersal vehicle that will give it protection and help it along. Thus, the egg cells of an apple tree are sent forth from the parent tree enclosed in apples, whereby each egg cell is surrounded by many thousands of protective cells forming the tough seed coat and delicious fruit flesh. Likewise, the queen bees of a honeybee colony are sent forth from the parent colony enclosed in swarms, whereby each queen is surrounded by some ten thousand worker bees providing a living shelter and food supply. Because each swarm or apple is many thousand times larger and more costly than each drone bee or pollen grain, it is no surprise that a colony or tree produces relatively few female units each year, usually fewer than four swarms and at most a few hundred apples. But because the costly female propagules are well protected and richly endowed, they have a high probability of successfully establishing a new colony or tree. So, despite their smaller numbers, swarms and apples match the effectiveness of drone bees and pollen grains in propagating the genes of their parents.

Swarming

In upstate New York, where I live, my colonies begin sending forth their drones in late April, and they begin casting their swarms—each one consisting of a queen accompanied by several thousand workers—a week or two later in early May. In essence, colony reproduction starts up shortly after winter shuts down. Most years, the swarming season begins after we've enjoyed a few weeks with warm days and profuse flowering by the maple trees (*Acer* spp.), pussy willow bushes (*Salix discolor*), and skunk cabbage plants (*Symplocarpus foetidus*). During this time, the colonies have collected much food, their queens have diligently laid eggs, and their worker populations have rapidly strengthened. I can predict with fair reliability when I will find my first swarm by noting when my hive of bees mounted on platform scales finally ends its six-month-long free-fall in weight and begins to bulk up again on fresh nectar and pollen (see fig. 2.8).

Swarming starts early in the summer because each new colony has much to accomplish if it is to survive the following winter. Specifically, each swarm (new colony) must locate a suitable nesting cavity, occupy it, and then build a set of beeswax combs, raise new workers, and store sufficient provisions to last through winter. Getting an early start certainly helps a colony clear these hurdles. Even so, sadly, many new colonies don't store up enough honey and so starve during their first winter. In the mid-1970s, for three years I followed the fates of several dozen feral honeybee colonies living in trees and houses around Ithaca, and I found that less than 25 percent of the "founder" colonies (ones newly started by swarms) would be alive the following spring. In contrast, almost 80 percent of the "established" colonies (ones already in residence for at least a year) would survive winter, no doubt because they hadn't had to start from scratch the previous summer. Beekeepers describe the time and energy crunch faced by swarms in a rather grim, three-line rhyme: "A swarm of bees in May is worth a load of hay, a swarm of bees in June is worth a silver spoon, a swarm in July isn't worth a fly."

Whether in May, June, or July, the first step that a colony takes to prepare for swarming is the rearing of 10 or more queens, all daughters of the mother queen. Queen rearing starts with the construction of queen cups, tiny inverted bowls made of beeswax. They are built usually along the lower edges of the combs in which the colony is producing brood and they will form the bases of the large, downward-pointing, peanut-shaped cells in which the queens will be reared (fig. 2.10). Next, the queen lays eggs in a dozen or more of the queen cups and workers feed the hatching larvae the royal jelly that ensures their development into queens. An enduring mystery about honeybees is what exactly stimulates a colony to begin rearing queens and thereby initiate the process of swarming. Beekeepers know that certain conditions inside a colony's hive (congestion of the adult bees, numerous immature bees, and expanding food reserves) and outside the hive (plentiful forage and spring time) are correlated with the start of queen rearing for swarming. Nevertheless, to this day, no one knows what specific stimuli the worker bees are sensing and integrating when they make the critical decision to start the swarming process.

The development of the new queens is remarkably rapid, requiring only 16 days from the time the egg is laid to the moment when an adult queen emerges from her cell. While these daughter queens develop, the mother queen undergoes changes that will prepare her for departure in the swarm. With each passing day, she is fed less and less by the workers. Her egg production declines, and her abdomen, no longer swollen with fully formed eggs, shrinks dramatically.

Fig. 2.10 One of the large, peanut-shaped cells in which queens are reared.

Fig. 2.11 A worker bee shaking the queen. The arrow indicates the dorso-ventral vibration of the bee's body.

Furthermore, the workers begin to show mild hostility toward their mother, shaking, pushing, and lightly biting her. Each time a worker shakes the queen, she grasps the queen with her forelegs and shakes her own body for a second or so, delivering 10 to 20 vigorous shakings of the queen (fig. 2.11). These bouts of rough handling, which eventually can become nearly continuous (occurring every 10 seconds or so), force the queen to keep walking about the nest. This increased exercise, together with the reduced feeding, results in a 25 percent reduction in the queen's body weight. In this way the mother queen, usually too large and heavy to fly, is put into flying trim.

While the daughter queens are maturing and the mother queen is slimming, the workers are also preparing for the impending mass departure of the mother queen and thousands of workers in a swarm. To ensure that they will be well supplied with energy when they leave home, the workers do just the opposite of slimming; they stuff themselves with honey, causing their abdomens to swell noticeably. A study in which the stomachs of workers from colonies preparing to swarm were painstakingly dissected and weighed found that most bees had filled their stomachs with a drop or two (35 to 55 milligrams) of honey, thereby increasing their body weights by about 50 percent. Thus when a swarm leaves on its journey to a new home, approximately one-third of its weight is a food reserve. The bloating of the workers is not their only conspicuous adjustment in anticipation of swarming. The wax glands, located on the ventral plates of four of the abdominal segments of each worker bee, become hypertrophied in preparation for the intense wax secretion needed for comb building at the new nest site. Turning over a worker bee plucked from a colony that is poised to swarm will reveal white scales of beeswax projecting from the overlapping ventral plates

Fig. 2.12 Wax scales on the undersides of the abdomens of worker bees.

(fig. 2.12). But what is perhaps the most striking change in the workers just before swarming is their greater lethargy. Many of these laggards hang quietly on the combs, while others rest in a thick cluster outside the hive entrance, giving the alert beekeeper a helpful warning that swarming is imminent. The biologist and comic book artist Jay Hosler has nicely called this period of odd inactivity "the calm before the swarm." Several dozen bees, however, remain active and start scouring the countryside for five or more kilometers (three miles) in all directions for possible nest sites. These enterprising individuals are nest-site scouts, the central figures in this book, and in chapter 4 we will see who they are.

In the summer of 2007, I learned that the nest-site scouts play a key role in triggering the next main event in swarming: the swarm's explosive departure from the parental nest. My partner in this work was one of my graduate students, Juliana Rangel, who is a fine scientist, being smart, cheerful, and hardworking. We learned that the scout bees are especially well qualified to instigate the swarm's mass exodus because their special occupation causes them to spend time both outside and inside the nest: outside to hunt for potential dwelling places, and inside to refuel and rest. Only a bee that has information from both indoors and outdoors can get the timing right for the swarm's departure. From her time inside, a scout bee can tell when some of the developing queens have reached the pupal stage and have had their cells sealed, and from her time outside, she can tell when the weather is sunny and warm, hence favorable for a journey. When both of these requirements for swarming are fulfilled, the scouts burst into action. Starting in the cluster of bees just outside the hive entrance, the excited scouts begin scrambling among their cool, calm sisters. Every few seconds, each scout will pause by a quiet bee and briefly press her thorax against the other bee while activating her flight muscles to produce a 200- to 250-hertz (cycles per second) vibration that lasts for a second or so. This signal is called worker piping. It sounds (because of high-frequency harmonics) like the engine of a Formula One race car making an all-out acceleration, and it informs the quiescent bees that it is time to warm their flight muscles by shivering to a flight-ready temperature of 35°C (95°F) in preparation for the swarm's departure. The piping by the scout bees is intermittent and faint at first, but over the next hour or so it gradually becomes steady and loud, as more and more of the scouts blast out the message "Time to warm up!" Ultimately, the piping-hot scout bees sense that all their hive-mates are flight ready—perhaps by consistently contacting suitably warmed bees—at which point the scouts start producing a second arousal signal, the buzz-run, in which each scout bee runs about the nest in great excitement, tracing out a crooked path, buzzing her wings in bursts, and bulldozing between sluggish bees. The message now is "Time to go!"

And go they do! Now nearly all of the worker bees become excited and run about, crowding toward the entrance opening where they pour out in a torrent and take to the air, pushing the mother queen out as well, creating what beekeepers call the "prime swarm" (fig. 2.13). It contains some ten thousand bees, about two-thirds of the colony's population. These swarming bees fly round each other in a wild whirl, forming a cloud approximately 10 to 20 meters (30 to 60 feet) across, with their queen flying somewhere in their midst. They don't go far. Soon, some of the workers settle on the branch of a tree or similar object, the queen

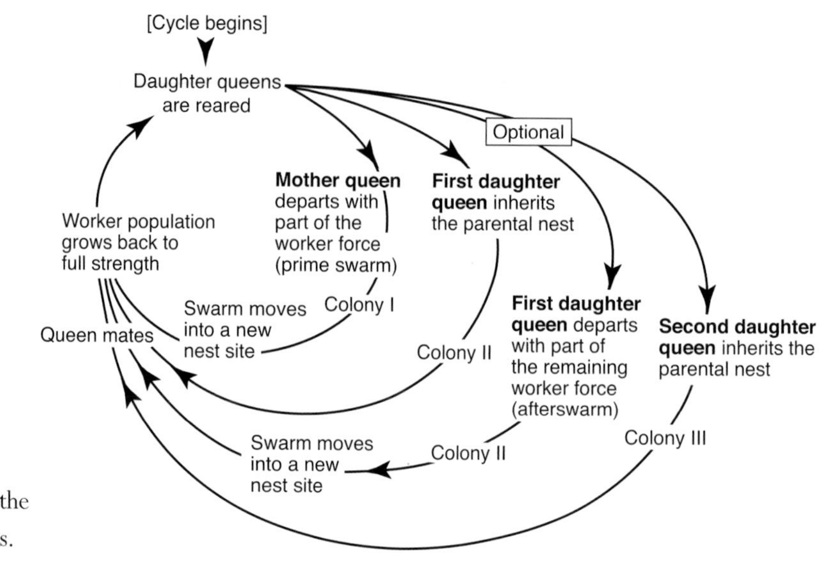

Fig. 2.13 Principal events in the life cycle of honeybee colonies.

joins them, and over the next 10 or 20 minutes the whole cloud of bees will condense into a beard-shaped cluster. The worker bees are attracted by the scent of the queen and by the strong lemony odor of the attraction pheromones that the first settlers are releasing from their scent organs (located near the tip of the abdomen) and are dispersing by fanning their wings. Over the next several hours, or several days, most of the swarm bees will hang here quietly, while the scouts will busily search the neighborhood for candidate dwelling places and choose a suitable abode. Once the scouts have completed their democratic decision making, they will induce the whole swarm to again take flight and then will guide the flying swarm to its new home.

Back in the parental nest, there remain a few thousand worker bees, a dozen or more queen cells, many thousand cells of worker brood, and much food. The stay-at-home workers are now without a queen, but not many days will pass before the first of the new queens emerges. During the waiting period, the parent colony's worker population will rebound as new workers emerge. Often, so many new workers appear that the colony's strength is restored by the time the first virgin queen emerges from her sealed queen cell. If the colony does regain its strength, then the workers will chase the first virgin queen away from the

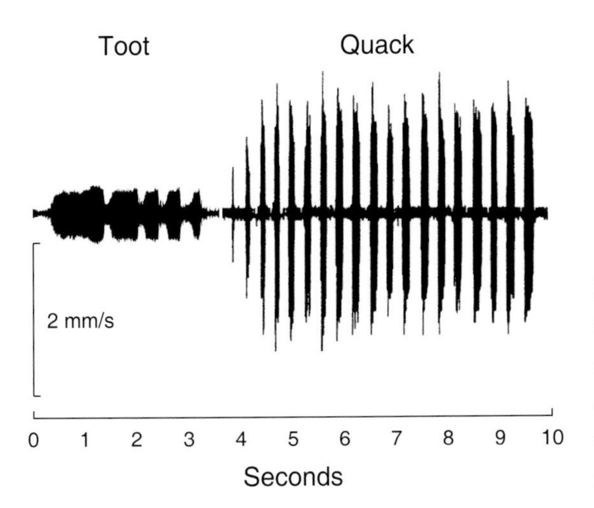

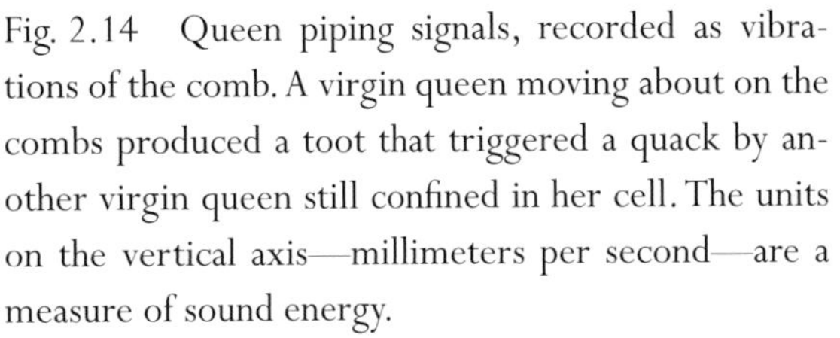
Fig. 2.14 Queen piping signals, recorded as vibrations of the comb. A virgin queen moving about on the combs produced a toot that triggered a quack by another virgin queen still confined in her cell. The units on the vertical axis—millimeters per second—are a measure of sound energy.

remaining queen cells to prevent her from destroying them. The workers will also refrain from chewing away the wax and pupal cocoon fibers on the capped ends of the queen cells to prevent the other virgin queens from getting free, and they will feed the confined queens whenever they beg for food by pushing their tongues through little slits in their cells. At the same time, the first virgin queen to emerge will announce her presence with queen piping signals, called "toots." A queen pipes the same way as a worker, by pressing her thorax against a substrate and activating her flight muscles. A queen, however, presses herself against a comb instead of against a bee, probably to give her signal a broader audience. Also, a queen's piping signals are longer than those of a worker, for they contain multiple pulses (fig. 2.14). When the first virgin queen pipes, the workers instantly cease all movement for the duration of her signal, perhaps to minimize the background noise produced by their myriad footsteps, and the virgin queens confined in their cells will pipe in response, producing lower-pitched "quacks" that are somewhat longer than the first virgin queen's "toots." These quacks almost certainly inform the first virgin queen that she has lethal rivals.

This bad news may encourage the first virgin queen to leave in a secondary swarm, what beekeepers call an "afterswarm." Doing so means that the first virgin queen relinquishes the wealth of desirable resources in the parent nest—the beeswax combs, worker brood, and honey stores—and starts down the risky path of founding a new colony. This course of action is, however, probably less dangerous for her than staying home and attempting to kill all her deadly serious competitors. Soon, the workers will start shaking the first virgin queen to prepare her for flight, and in a few days, if good weather prevails, they will push her out of the nest during the departure of a second swarm. This process is repeated with each emerging queen until the colony is weakened to the point where it cannot support further swarming. At this point, if there are still multiple virgin queens in the nest, the workers will allow them to emerge freely. The first one out usually attempts to kill those still in their cells by dashing over the combs in search of cells containing queens, chewing small holes in their sides, and stinging the occupants. If, however, two or more virgin queens emerge together, they will fight to the death, seizing each other and attempting to sting. The battling queen bees grapple and twist, each one struggling fiercely to implant her venom-laden sting in her sister's abdomen. Ultimately, one queen succeeds and the other, fatally stricken, collapses in paralysis, falls from the comb, and soon dies. The merciless sororocide continues until just one virgin queen remains alive. Several days later the victor will make her mating flights and, once fully mated, start her egg laying. Soon her daughters and sons will populate the coveted parental nest. Any virgin queen who departed in an afterswarm will likewise make mating flights once she and her workers have moved into their new home, for no queen ever mates inside a nest.

3
DREAM HOME FOR HONEYBEES

If I can with confidence say
That still for another day,
Or even another year,
I will be there for you, my dear,

It will be because, though small
As measured against the All,
I have been so instinctively thorough
About my crevice and burrow.
—*Robert Frost,* A Drumlin Woodchuck, *1936*

Like Robert Frost's woodchuck, a honeybee colony is "instinctively thorough" about its dwelling place, for only certain tree cavities provide good protection from predators and sufficient refuge from harsh physical conditions, especially strong winds and deep cold. No fewer than six distinct properties of a potential homesite—including cavity volume, entrance height, entrance size, and presence of combs from an earlier colony—are assessed to produce an overall judgment of a site's quality. The care with which honeybees choose their homes has been known for only about 30 years, which might seem surprising given that humans have been culturing these bees since ancient times. The reason that humans have only recently learned about the bees' real estate preferences is that the essence of beekeeping is the tending of colonies living in hives fashioned by a beekeeper and sited where the beekeeper wants them. The earliest solid evidence of bee-keeping comes from Egypt, around 2400 BC, and consists of a stone bas-relief in

Fig. 3.2 Bee tree, with a knothole that serves as the nest entrance visible high up in the left fork.

these investigations were all conducted indoors in clean, brightly illuminated, and nearly lifeless laboratories. But now, as a beginning graduate student in biology and novice investigator of animal behavior, I was keen to work outdoors using what has been called the von Frisch–Lindauer approach to animal behavior research. In their autobiographical book, *Journey to the Ants*, Bert Hölldobler and Edward O. Wilson explain that von Frisch and Lindauer had a philosophy of research based on:

> a thorough, loving interest in—a feel for—the organism, especially as it fits into the natural environment. Learn the species of your choice every way you can, this whole-organismic approach stipulates. Try to understand, or at the very least try to imagine, how its behavior and physiology adapt it to the real world. Then select a piece of behavior that can be separated and analyzed as though it were a bit of anatomy. Having identified a phenomenon to call your own, press the investigation in the most promising direction.

My thesis advisor, Bert Hölldobler, had presented this way of studying behavior in his ethology course at Harvard and, more importantly, had demonstrated its power by his own spectacularly beautiful studies of ant social behavior. So, by the end of my first year in graduate school, I was raring to go. I wanted to gain a feel for honeybees living in nature, to further analyze the house-hunting piece of their behavior, and to see if I could press the investigation on from where Martin Lindauer had left it some 20 years before.

I knew that I would abscond from Harvard the moment I had finished taking my final exams for the spring semester, and I had my mind set on returning to the Dyce Laboratory for Honey Bee Studies, at Cornell, where I had worked for the previous four summers when an undergraduate student. The director of the lab, Professor Roger A. Morse, was truly a generous man. He welcomed me back, assigned me a desk, and provided several essential tools for the project—a powerful chain saw, steel wedges and maul, and one of the lab's green pickup trucks. Most importantly, "Doc" Morse arranged for me to team up with a member of the Entomology Department's technical staff, Herb Nelson, who had worked as a logger in the Maine woods when a teenager and could teach me how to cut down big trees without getting killed.

Herb and I started with some of the bee trees I had discovered back in high school while exploring the woods around my family's home. These were augmented with ones that I located through a want ad I placed in the local newspaper, the *Ithaca Journal*. The ad read, "**BEE TREES** wanted. Will pay $15 or 15 lb of honey for a tree housing a live colony of honeybees. 607-254-5443." I feared I'd get no calls, but within a week I had secured the rights to 18 accessible bee trees in the woods around Ithaca. Two owners took payment in money; all the rest wanted honey.

The procedure for collecting these nests was simple but somewhat dangerous. Shortly before sunrise, when all bees were still at home, I would hike to a bee tree with a can of calcium cyanide powder (Cyanogas), an old spoon, and some rags. If the nest entrance was high in a tree that I couldn't climb, I'd also bring an aluminum extension ladder. My aim was to spoon cyanide powder into the nest entrance and then quickly plug it with the rags. The cyanide powder would react with moisture in the air producing cyanide gas that would kill the bees but, if all went according to plan, not me. (Once I did drop the can of Cyanogas from the ladder, spilling much of its contents, but I managed to hold my breath long enough to climb down, get the lid back on the can, and dash out of the expanding cloud of deadly gas.) By first killing the bees, we could later fell the tree and collect the nest without being ferociously attacked. This protocol also enabled me to census the bee population of each wild colony when I dissected its nest.

Having killed the bees, I'd return to Dyce Lab to pick up Herb and load the truck with the tools we'd need for the day: chain saw, wedges and maul, rope, ramp boards, tape measure, magnetic compass, 35 mm camera, and notebook. Our goal was to cut down the bee tree I had just visited, saw out the trunk section housing the nest, wrestle it onto the truck, and haul it back to the lab. I recall being impressed by Herb's confidence in driving the truck deep into the woods to get near each bee tree ("We'll have plenty of traction for getting back out, once we get that big log loaded on.") and by his careful inspection of each tree's lean and crown before starting his cutting ("You gotta know which way the tree wants to fall."). Herb's lumberjack skills weren't rusty, and each tree arced down neatly into the woods opening he had chosen. Once we had a tree lying on the ground, we proceeded to cut out the section containing the nest. We did this by making a series of crosscuts, starting far above and far below the nest entrance, and then working our way closer and closer to the entrance until the chain saw started spitting dark-brown punk wood or yellow-brown beeswax, indicating we had breached the nest cavity. We then rolled the nest-containing log—sometimes a massive, 2-meter-long (6-foot) and nearly 1-meter-thick (3-foot) section of the tree's bole—up into the truck, got it back to the lab, and split it open (fig. 3.3). Finally, we would lug the opened nest indoors where I could dissect it carefully under good light while measuring important features of the nest cavity and its

Fig. 3.3 Natural honeybee nest in the bee tree shown in figure 3.2. The tree section housing the nest has been split open, revealing the combs containing honey (above) and brood (below). The entrance hole is on the left side, about two-thirds of the way up the cavity.

contents. To measure the volume of the cavity, I filled it with liter after liter of sand after removing the combs. As I picked through the broken combs and dead bees, sooner or later coming across the lifeless queen, I felt sad to have killed a whole colony, but also excited, knowing that I was the first human to describe in detail the natural homes of honeybees.

Over the summer of 1975, we collected and I dissected 21 bee tree nests, enough to give us a broad picture of the nests of wild colonies living in the woods. I also located another 18 nests in trees that were left standing and so yielded information only about their entrance openings. Since the nest entrance is the "front door" of a colony's home it is probably especially important to the bees, so

I gave it extra attention. We found that the bees occupied many kinds of trees, including oaks (*Quercus* spp.), walnuts (*Juglans* spp.), elms (*Ulmus* spp.), pines (*Pinus* spp.), hickories (*Carya* spp.), ashes (*Fraxinus* spp.), and maples (*Acer* spp.). This suggested that the bees don't have a strong preference for certain tree species.

Not surprisingly, the tree cavities occupied by the bees were generally tall and cylindrical, consistent with the shape of tree trunks. But what was surprising was the discovery that most of these wild colonies were occupying tree cavities much smaller than the hives provided by beekeepers. The average nest cavity was only about 20 centimeters (8 inches) in diameter and 150 centimeters (60 inches) tall; hence, it had a volume of only about 45 liters (41 quarts) (fig. 3.4). A tree cavity of this size provides only one-quarter to one-half of the living space provided by a beekeeper's hive. Were the bees telling me that they prefer relatively small and snug nest sites, ones in which it might be easier to keep warm in winter? Some of the colonies even occupied tree cavities with only 20 to 30 liters of nesting space, though none was found in a space smaller than 12 liters. Was this lower limit of about 12 liters a sign that bees carefully avoid excessively cramped quarters, ones lacking sufficient room for storing the honey needed to survive winter? Certainly the bees living in these tree cavities were making good use of their living space, for each colony had nearly filled its nest cavity with multiple combs. Because each comb formed a wall-to-wall curtain spanning the (generally) narrow tree cavity, I was impressed by the way the bees had built small passageways in the combs where they were attached to the cavity's wall, so they could crawl easily from one comb to the next. And it was clear that these bees had organized their use of their combs in the way familiar to all beekeepers, storing honey in the upper region of the nest and rearing brood below. The nests collected in August, by the way, revealed that most colonies had been making good progress in stockpiling their winter heating fuel. The nests that I dissected contained, on average, 14 kilograms (30 pounds) of golden honey. Regrettably, it was all laced with cyanide.

The entrance openings of the bees' nests also showed consistencies that suggested possible nest-site preferences by the bees. Most nest entrances consisted of a single knothole or crack with a total area of just 10 to 30 square centimeters (2 to 5 square inches) (fig. 3. 5). And typically they were located near the floor of the tall tree cavity, on the south side of the tree, and close to ground level.

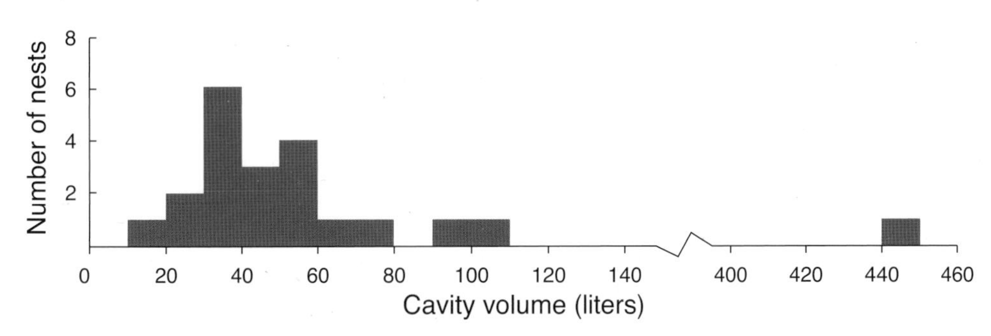

Fig. 3.4 Distribution of nest-cavity volumes for 21 nests in hollow trees.

The trends toward small size, floor level, and southern orientation all made good sense to me, for they would make the nest cavity inaccessible to most predators, relatively free of drafts, and perhaps warmed by the sun—all things that would be good for a colony. But the preponderance of nest entrances just a few feet from the ground puzzled me greatly. I figured that a low nest entrance must render a colony vulnerable to detection by predators, such as bears, whose attacks can be fatal. And I knew that in medieval times, in the forests of northern Europe (Germany, Poland, and Russia), one of the ancestral homes of the honeybees imported to North America, raids by bears on honeybee nests in trees were such a great trouble for the forest beekeepers who owned these nests that they invented horrific devices to kill honey-loving bears. One was a hinged platform mounted outside a bee nest. When a bear climbed onto it to attack the bees, it would collapse, causing the bear to tumble onto a grid of deadly sharp stakes below.

So at first I was perplexed by the rarity of nests high in trees. But as will be explained shortly, we now know that bees actually have a strong preference for nesting cavities with entrances located high above the ground. I also now know that my initial report of most nests being near ground level was an error generated by an unintentional bias in the way I had sampled the population of natural nests. Because the nests I studied were ones that had been noticed inadvertently by a person walking past a bee tree, and because people are much more likely to notice bees trafficking from a ground-level nest entrance than a tree-top one, I unwittingly studied nests whose entrances were far lower than is typical. I am confident on this point because several years later, when I became a bee hunter

Fig. 3.5 Knothole entrance of the nest in the bee tree shown in figure 3.2, showing some of the bees inside. This entrance opening is approximately 5 centimeters (2 inches) wide and 8 centimeters (3 inches) tall.

and mastered the ancient craft of lining bees (locating bee trees by baiting foragers from flowers and observing their flights back to their nests), I found that every hunt ended with me straining to spy the bees zipping in and out of a nest entrance high in a tree, like the one shown in figure 3.2. To date, I have located 27 bee trees by bee lining and can report that the average height of their nest entrances is 6.5 meters (21 feet). Needless to say, I'm now alert to the hidden danger of unintentional sampling bias.

Location, Location, Location

Even though the descriptive study of the natural homes of honeybees was destructive, it remains one of my favorite studies, for it put me in touch with honeybees living in nature and it helped me gain some self-confidence as a researcher. It also guided me throughout the next step of my investigation of how honeybee swarms choose a home: testing whether the nest-site patterns we had found—in cavity volume, entrance area, entrance height, and so forth—were a result of preferences of scout bees or were simply a consequence of the tree cavities that were available. The idea for the design of the test came from what I had read about beekeeping in East Africa and South Africa. In these regions, beekeepers acquire bees by hanging hives (usually hollowed logs with the ends stoppered except for an entrance hole) in trees and waiting for swarms to occupy them. I had not heard or read of anybody in North America catching swarms with "bait hives," but I reasoned that if one could do so, then I could ask the bees about their nest-site preferences by putting up groups of two or three nest boxes, with the boxes in each group identical except for one property, such as cavity volume or the height of the entrance above the ground. I hoped that scouts from wild swarms would discover the groups of nest boxes and would reveal their real-estate preferences by choosing among the boxes in a group and consistently occupying those with certain attributes.

Almost always, one starts an experimental study with a small-scale, low-cost pilot study to figure out what methods are apt to work before undertaking a costly full-scale investigation. In the summer of 1975, I did a pilot study to see if wild swarms would occupy bait hives often enough to give my experimental plan a reasonable chance of success. Using some scrap plywood scrounged from Dyce Lab, I built six nest boxes, each one a simple cubic box 35 centimeters (14 inches) wide, tall, and deep, and bearing a 4.5-centimeter (1.75-inch) diameter entrance hole on the front side. I designed these boxes to mimic the nest cavities I was finding in bee trees. My bee houses looked like birdhouses on steroids, except that each one had chicken wire nailed over its entrance opening to keep birds out while letting bees in. I took each nest box to a place I knew that I'd enjoy visiting in my home "territory" of Ellis Hollow, and nailed it about 5 meters (15 feet) off the ground onto the side of a large tree. I still remember vividly the thrill I felt a few weeks later, in late June, when I checked the nest box that I'd mounted on a dead elm tree along Cascadilla Creek and saw dozens of leather-colored honeybees bustling in and out of its entrance. A swarm had moved in! Yippee!! When swarms occupied two more of my nest boxes over the next few weeks, I was even more excited. This pilot study had been simple, but it had yielded the triumph of knowing that my experimental plan was likely to work. My plan for the next summer was now clear: I had to set out dozens and dozens of nest boxes of various designs to "ask the bees about this matter" of their ideal dwelling place.

The plan worked well. Each summer in 1976 and 1977, I set up more than two hundred green nest boxes in groups of two or three across Tompkins County and each summer over half of my nest-box groups attracted at least one wild swarm. The boxes within each group were spaced about 10 meters (33 feet) apart on similar-sized trees or, even better yet, on power-line poles where they were perfectly matched in visibility, wind exposure, and the like (fig. 3.6). Each group of boxes was designed to test one nest-site preference, and it did so by giving swarms a choice between one box whose properties all matched those of a *typical* nest site in nature (e.g., average entrance area, average cavity volume, etc.) and one or two other boxes identical to the first except in one property, the value of which was *atypical*. In this way, wild swarms were tested for a preference in the one variable that differed between the boxes. For example, to test for a preference in entrance size, I set up pairs of cubical nest boxes that were identical except that one had a typical entrance area of 12.5 square centimeters (2.5 square inches) and the other had a larger than usual entrance area of 75 square centimeters (15 square inches). Similarly, to test for a preference in cavity size, I set up trios of cubical nest boxes that were identical except that one box had the typical cavity volume of 40 liters while the other two boxes had volumes at the two tails of the distribution of nest cavity volumes: 10 and 100 liters.

To build the many nest boxes needed for this study, I spent most of my Christmas break in 1975 sawing and hammering and painting in the woodshop at Dyce Lab. There I constructed 252 nest boxes and used up enough plywood (more than 70 sheets) for building a small house. With these hundreds of nest boxes, I would eventually capture 124 swarms in 1976 and 1977.

As is shown in table 3.1, the swarms demonstrated preferences in the following nest-site variables: entrance size, entrance direction, entrance height above the ground, entrance height above the cavity floor, cavity volume, and presence of combs in the cavity. The bees had revealed to me that they prefer a nest entrance that is rather small, faces south, is high off the ground, and opens into the bot-

Fig. 3.6 Two nest boxes mounted on power line poles. The two boxes offer identical nesting sites (same cavity volume and shape, same entrance height and direction, etc.), except that the one on the right has a smaller entrance opening (12.5 square centimeters or 2 square inches) than the one on the left (75 square centimeters or 12 square inches).

tom of the nest cavity. These four preferences regarding the entrance opening no doubt help a honeybee colony survive against threats of cold winters and dangerous predators. A small entrance is easily defended and helps isolate the nest from the outside environment. An entrance high up in a tree is less apt to be discovered by predators than one near the ground, and is certainly inaccessible to predators that cannot fly or climb trees. An entrance at the bottom of the nest cavity rather than at the top may help to minimize the loss of heat from the colony by convection currents. And an entrance that faces south provides a warm, solar-heated porch from which foragers can take off and on which they can land. Beekeepers, incidentally, face their hives to the south to help their bees fly out in cool weather. This southern orientation is particularly important in the winter months, when bees go outside on sunny days to make their critical "cleansing flights," that is, to defecate. A Canadian bee researcher based in Alberta, Tibor Szabo, compared colonies living in south-facing and north-facing hives. He found that those in south-facing hives were less apt to suffer a hive entrance plugged by ice in winter and were more populous in spring.

The pattern of nest-box occupations by swarms also showed clearly that the bees avoid cavities smaller than 10 liters or greater than 100 liters, and that they very

Table 3.1
Nest-site properties for which honeybees do or do not show preferences, based on nest-box occupations by swarms.

Property	*Preference*	*Function*
Size of entrance	12.5 > 75 cm^2	Colony defense and thermoregulation
Direction of entrance	South > north facing	Colony thermoregulation
Height of entrance	5 > 1 m	Colony defense
Position of entrance	Bottom > top of cavity	Colony thermoregulation
Shape of entrance	Circle = vertical slit	None
Volume of cavity	10 < 40 > 100 liters	Storage space for honey and colony thermoregulation
Combs in cavity	With > without	Economy in nest construction
Shape of cavity	Cubical = tall	None
Dryness of cavity	Wet = dry	Bees can waterproof a leaky cavity
Draftiness of cavity	Drafty = tight	Bees can caulk cracks and holes

A > B, denotes A is preferred to B; A = B denotes no preference between A and B.

much like 40-liter cavities (about the size of a wastebasket), especially ones already equipped with combs. Probably the main problem the bees face regarding cavity volume is avoiding undersized cavities, since most tree cavities are too small (less than about 15 liters) to hold the store of honey a colony needs to survive winter. The evidence supporting this assertion comes from a small study I did with one of my brothers, Daniel H. Seeley, who owns a whole hillside in Vermont that was logged in the 1800s but is now forested with stately sugar maple (*Acer saccharum*) and beech (*Fagus grandifolia*) trees. In October 1976, Dan and I packed up logging tools and drove north from Cambridge, Massachusetts, to Roxbury, Vermont, to spend several days of Indian summer finding out what size cavities scout bees are apt to encounter when prospecting for homesites. Working over a 0.32-hectare (0.8-acre) area, we felled every tree more than 30 centimeters (12 inches) in diameter, and we sawed each felled tree into 120-centimeter (4-foot) lengths to expose any cavities they contained. We dissected 39 trees and found 14 cavities with an opening to the outside that could provide access to a scout bee. Of these 14 tree hollows, only two (14 percent) were larger than 15 liters; they were 32 and 39 liters.

The preference for a site filled with combs—built by a preceding colony that did not survive winter—doubtless reflects the tremendous energy savings that a colony enjoys if it occupies a site already furnished with a full set of combs. The energy thus saved turns out to be a large fraction of the honey store that a fledgling colony needs to survive its first winter. This is shown by the following calculations. A typical nest in a bee tree contains some 100,000 cells arranged in eight or so combs whose total surface area is about two and half square meters (3 square yards). Building this impressive edifice requires about 1,200 grams (2.5 pounds) of beeswax. Given that the weight-to-weight efficiency of beeswax synthesis from sugar is at most about 0.20, we can estimate that building the combs in a typical nest requires about 6.0 kilograms of sugar, hence about 7.5 kilograms (16 pounds) of honey. This mass of honey is about one-third of what a colony will consume over winter. Storing these 7.5 kilograms of honey in the colony's food supply for winter rather than spending it on comb building will greatly boost a colony's odds of surviving its first winter. Recall that I found that 76 percent of the colonies newly established in tree cavities around Ithaca die during their first winter, and that nearly all of the colonies that do so succumb to starvation.

The nest-site properties for which I detected no preference were entrance shape, cavity shape, cavity draftiness, and cavity dryness. Honeybees probably prefer tight and dry nest cavities, but because a colony can caulk with tree resins any cracks and crevices that let in drafts and water, the nest-site scouts apparently do not pay much heed to these properties. In contrast, a colony cannot modify the volume of a nest cavity, the height of its entrance, or the direction in which it faces, so to get a homesite that meets its needs in these properties the nest-site scouts must pay close attention to these properties when evaluating prospective nest sites. The ability of a honeybee colony to remedy a drafty or damp site was neatly demonstrated by the swarm bees that occupied my experimental nest boxes. I had rendered some of the boxes drafty by riddling their fronts and sides with six-millimeter (quarter-inch) diameter holes spaced 7.5 centimeters (3 inches) apart (fig. 3.7). Other boxes I had made damp by dumping 2 liters (2 quarts) of waterlogged sawdust onto the floor of each box. Every swarm that moved into one of the drafty boxes soon made it draft free by plugging with tree resins all the holes I had drilled. Likewise, every swarm that occupied one of the

Fig. 3.7 Two of the nest boxes used to test whether bees prefer a nesting cavity that is tight (right) to one that is drafty (left), with lots of holes in the walls.

damp boxes quickly rendered it dry by hauling out all the soggy sawdust that I had dumped inside. I was greatly impressed by the bees' tidiness.

Freebies

One thing that makes studying honeybees so enjoyable is the way that what is learned through curiosity-driven research often turns out, unexpectedly, to have real practical value. My best example of this phenomenon is the way that knowing something about the defecation habits of Asian honeybees helped defuse tensions between the United States and the Soviet Union back in the 1980s. This story starts in the late 1970s when I had finished graduate school and was keen to travel overseas and learn about the marvelous species of honeybees that live in the Asian tropics: the Asian hive honeybee (*Apis cerana*), the dwarf honeybee (*Apis florea*), and the giant honeybee (*Apis dorsata*). With support from the National Geographic Society, my wife Robin and I undertook a 10-month study of the colony defense strategies of the three Asian honeybee species living in Thailand. We set up camp in the pristine mountain forests of the vast Khao Yai National Park in northeast Thailand, where one can still enjoy the sight of hornbills winging their way between towering dipterocarp trees, the eerie smell of Asian tiger urine deposited along a trail, the whooping calls of white-faced gibbons shortly after sunrise, and the mysterious biology of the Asian honeybees. Gradually we assembled a picture of each honeybee species' fascinatingly complex array of colony defenses against such enemies as giant hornets, weaver ants, honey buzzards, tree shrews, rhesus monkeys, and honey bears. This was field biology done for biology's sake, and it was a wonderful adventure for two newlyweds. Sometimes I wonder, though, if even half a dozen biologists worldwide have read closely the beautifully detailed, 21-page report on the Asian honeybees that we published in the scientific journal *Ecological Monographs*.

A few years later, however, and to my amazement, the knowledge that we'd gained about the Asian honeybees proved important to a large international audience. In 1981, the secretary of state in the Reagan administration, Alexander M. Haig, alleged that the Soviet Union was waging or abetting chemical warfare against opponents of the communist governments in two countries bordering Thailand: Laos and Kampuchea. If true, this was a violation of two international arms-control treaties, the 1925 Geneva Protocol and the 1972 Biological Weapons Convention. The main evidence cited by Haig was a material called "yellow rain," that is, yellow spots less than 6 millimeters (one-quarter inch) in diameter that were found on vegetation at alleged attack sites and that supposedly contained fungal toxins. I realized, however, that the yellow spots that U.S. officials called yellow rain were indistinguishable from the yellow spots I called honeybee feces. They were identical in size, shape, and color. Further work revealed that both contained bee hairs and were laden with pollen grains from which the protein had been digested. Eventually, I was able to help Matthew Meselson, a professor of molecular genetics at Harvard and an expert on chemical and biological weapons, show conclusively that yellow rain was indeed honeybee feces, not chemical warfare. One wag said we had uncovered the work of "KGBs." Shortly after yellow rain was proven to be bee poop, in 1984, officials of the U.S. State Department, without fanfare, ceased accusing the Soviets of violating the two arms-control treaties on chemical and biological weapons.

The yellow rain story is a striking example of how research rooted in sheer curiosity can unexpectedly yield useful knowledge, but it is not so unusual, for real-world benefits often bubble up from basic research. My first taste of pursuing personal curiosity and getting an unexpected bonus of practical results came from my study with Doc Morse of the nest-site preferences of swarms. In the summer of 1976, we had groups of nest boxes set up in more than 100 sites around Tompkins County, and we caught more than 60 swarms. Given our high success rate in trapping swarms, Doc realized that we should translate our findings about the bees' nest-site preferences into recommendations for beekeepers on how to build and position bait hives to catch wild swarms of honeybees. We prepared a simple design (fig. 3.8) along with a set of guidelines for situating bait hives—a good location is about 5 meters (15 feet) off the ground, highly visible but fully shaded, and facing south—and published these in the beekeeping magazine *Gleanings in Bee Culture* and as a Cornell Cooperative Extension Bulletin. Beekeepers responded enthusiastically. Before this, beekeepers wanting to capture wild swarms had to rely on being notified when a swarm had settled somewhere, then they had to hurry to put it in a hive before the bees finished selecting a nest site and flew away to their chosen home. With bait hives, beekeepers can collect swarms automatically.

In recent years, other bee scientists have designed cheaper, lighter, and tougher bait hives made of reinforced wood pulp material, and have devised scent lures that slowly leak a 1:1:1 blend of chemicals (citral:geraniol:nerolic+geranic acids) from a small polyethylene vial. These scent lures mimic the attraction pheromones that scout bees release from their scent organs to mark a desirable homesite (discussed further in chapter 8). Experiments by Justin Schmidt at the USDA Bee Research Center in Tucson, Arizona, have shown that a bait hive with a scent lure can be five times more likely to attract a swarm than one without a lure, probably because the artificial attraction pheromones makes a bait hive much more likely to be discovered by a scout bee, but perhaps by also making it more attractive. Wood pulp bait hives (sometimes called "swarm traps") and swarm scent lures are now produced commercially and are sold by the companies that sell beekeeping equipment. Each summer, I put up a half dozen bait hives, partly because I can always use a few additional colonies of bees, but mostly because I like getting free bees.

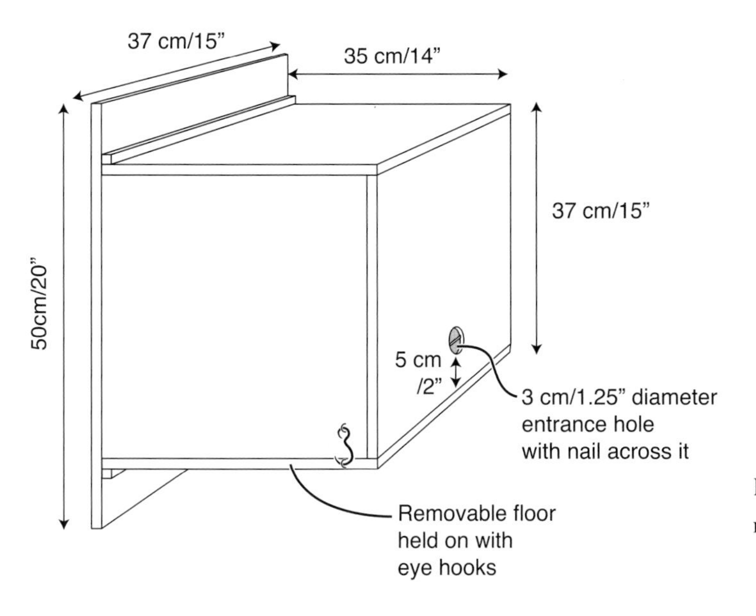

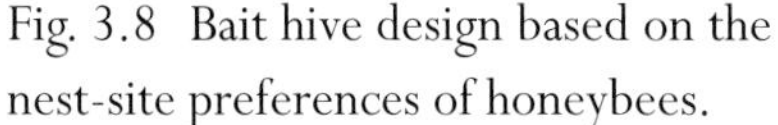
Fig. 3.8 Bait hive design based on the nest-site preferences of honeybees.

Property Assessments

Probably every homeowner has wondered how his or her local tax assessor combines information about the size of a house lot, the floor area of the house on the lot, the number of bedrooms and bathrooms in the house, and so forth to determine the assessed value of a particular property. I began wondering the same thing about scout bees in August 1974, as I watched several scouts scrutinizing a candidate dwelling place. This happened in the summer before I went off to start graduate school at Harvard. I was working happily for Doc Morse at Cornell's Dyce Laboratory, but I was a little worried about my choice of problem for a doctoral thesis: deepening Martin Lindauer's work on how a swarm chooses its home. In 20 years, nobody had tackled, much less solved, the many mysteries raised by Lindauer's study. Clearly, there was a first-rate opportunity here, but could I succeed in making something of it? To begin to see what I might do, I decided simply to watch, with my eyes wide open, a swarm go through its democratic decision-making process. In working for Doc, I had learned how to make an artificial swarm—by shaking a colony (queen and workers) into a cage to render them homeless, and then feeding them lavishly with sugar syrup to get them stuffed with food like natural swarm bees—so I prepared a swarm and set it up behind my parent's house in Ellis Hollow (see fig. 1.7). I also nailed a nest box that I built from scrap plywood to a white pine tree about 150 meters (500 feet) away, hoping the swarm's scouts would discover it and select it for their new homesite. I mounted the box at eye level so I could observe easily any scout bees that might visit it.

The weekend that I watched this swarm turned out to be a milestone in my life. The swarm's scouts quickly began advertising several prospective homesites with their dances, and before long one bee was dancing with eye-catching enthusiasm for a location nearby and to the north: my nest box! This bee's lively dance soon gave rise to a small crowd of bees at the box. Back at the swarm, I adorned a few of the bees dancing for my box with dots of paint on the thorax and abdomen, giving different bees different color combinations. This simple trick transformed these individuals from mere members of *Apis mellifera* into personal acquaintances whose affairs became of the greatest interest to me. At the swarm, I saw how a scout would perform a bout of vigorous dancing, contact another bee with agile antennal movements to beg a droplet of honey, perhaps groom off a pesky fleck of paint, and then fly away for 20 to 30 minutes. When she returned, she might dance again but she might also just settle quietly within the swarm cluster. At the nest box, I saw how my labeled bees would land and agitatedly run into the entrance opening and then scurry back out a minute or so later, whereupon they would either crawl briefly around the entrance and then pop back inside or conduct a slow, hovering flight around the box, usually facing it and maneuvering within inches of the box, apparently giving the nest structure a detailed visual inspection (fig. 3.9). Never before had I seen a honeybee behave with such persistent intention to gather information. I was thoroughly intrigued. Also, my worries about choice of thesis topic had largely vanished, for I felt confident that I could explore how a scout bee inspects a site.

I began to study closely the inspection behavior of nest-site scouts the following summer, in June 1975. To do so, I had to leave the forested countryside around Ithaca, where since early May I had been busy finding bee trees and describing natural nests, and shift to a location largely devoid of natural homes for

Fig. 3.9 Scout bees inspecting the front of a nest box.

honeybees: Appledore Island (fig. 3.10). This rocky, wind-blown island is barely 900 meters (half a mile) long and lies 10 kilometers (6 miles) out in the Atlantic Ocean off the coast of southern Maine. It is the site of the Shoals Marine Laboratory of Cornell University and the University of New Hampshire. I was attracted to Appledore because it has no resident honeybees and it has no large trees. Instead, it is inhabited by approximately one thousand breeding pairs of herring gulls (*Larus argentatus*) and great black-backed gulls (*Larus marinus*), and it is covered by thickets of poison ivy plants rising three meters (10 feet) tall together with tangles of blackberry brambles and wind-battered cherry bushes, all richly fertilized by the gulls. I figured that if I took a honeybee swarm out to this scrubby island, the bees would have to concentrate their house hunting on the artificial nesting sites that I would provide. If so, then I could observe their behavior under controlled conditions and learn how they assess a possible dwelling place.

My first goal on Appledore Island was to make a detailed description of the behavior of scout bees inspecting nest sites. I hoped these observations would suggest how scouts evaluate the critical nest-site properties. It was particularly important to be able to watch scouts inside a prospective homesite in order to understand how they might measure such things as the cavity's volume and the entrance's height from the cavity floor. To this end, I built a lightproof hut with a

Fig. 3.10 Top: Appledore Island, Maine, in the foreground, with several of the other eight islands in the Isles of Shoals archipelago in the background. Bottom: rocks, scrubby vegetation, and gull residents of Appledore, on a foggy morning.

Fig. 3.11 Interior of the nest box designed for observing a scout bee as she inspected a potential nest site. The box was mounted outside a red-filter window on the side of a lightproof hut. Because bees do not see red light, the bees could be watched and their movements recorded (by reference to the numbered squares) without disturbance.

cube-shaped nest box mounted on one of its walls (fig. 3.11). The box was positioned outside a window covered with a red filter (bees cannot see red light) so that I could peer inside the box without disturbing the scouts. The inner surfaces of the box bore a grid-coordinate system that enabled me to record where a scout bee went while she was in the cavity. After setting up the hut in a valley on one side of the island, I positioned a small swarm (about 2,000 bees) at the center of the island. The bees had been given dots of paint using a color code that made them individually identifiable. Then I retired to the hut to wait for scout bees.

The first time I tried this, I waited all morning at the hut without being visited by a scout bee, which surprised and dismayed me. When I returned to the swarm around midday, my spirits sank further, for I saw several scouts performing long-lasting dances indicating a location directly away from my hut. Rats! What possibly could the bees have found? I made careful measurements of the direction and distance to the location specified by the bees' dances and plotted the site on my

Fig. 3.12 Island cottage of Rodney Sullivan, lobster fisherman. Ladder resting against roof shows how the author reached the chimney to screen off the bees.

topographical map. Now my spirits sank even deeper, for the bees' dances were indicating unmistakably one of the two lobster fishermen's cottages on the south shore of the island, specifically Rodney Sullivan's cottage (fig. 3.12). When I had arrived on Appledore a few days before, and had been getting oriented to my new surroundings, I had been told to keep well away from the fishermen's private properties, especially Rodney's place, for he valued his privacy and kept a loaded shotgun behind his front door. What should I do? I sought advice from the laboratory's director, Professor John M. Kingsbury, and he kindly went with me over to Rodney's cottage, to introduce student Seeley to fisherman Sullivan. We went by boat so Rodney would be able to see us coming, approaching from the front (the water), even while my bees were "attacking" from the rear (the land). He heard our boat approach, came out on his porch, and told us to come ashore. After we climbed up the rocks to his house, he told us he had an emergency: hundreds of bees were buzzing in the stovepipe to his woodstove! "Never seen this before!

Could they'a been blown out here during the [recent] storm?" I didn't answer the question, but I did offer to help. While Rodney made a fire in the stove to smoke out the bees, I climbed his steep roof (slippery with splotches of fresh gull poop) and taped window screen over his chimney flue to exclude bees in the future. Rodney was delighted . . . I was relieved.

No longer distracted by Rodney's house, the scouts from my swarm soon started to appear at my observation nest box and began to perform their striking inspection behavior. I saw that a scout bee needs 13 to 56 minutes (average 37 min) to inspect a prospective nest site. Her complete inspection is a summation of 10 to 30 journeys inside the cavity, each one lasting usually less than a minute and alternating with equally brief periods outside, during which the bee examines the exterior of the nest structure. I call this first inspection, when a scout is popping in and out of the cavity, the discovery inspection. Following the discovery inspection, a scout returns to the swarm, and if the site is desirable she will advertise it with a waggle dance and usually will then make repeated visits to the site at approximately half-hour intervals, but these subsequent site visits tend to last only 10 to 20 minutes (average 13 min) and don't involve so much in-and-out activity.

When a scout is inside a cavity, conducting her discovery inspection, she devotes most of her time (about 75 percent) to rapid walking across the inner surfaces. This quick pacing about is interspersed with pauses to rest, groom, and release attraction pheromones from the scent organ, and with short hopping flights. The inside of a dark cavity seems an odd place to attempt to fly about, yet the bees make these little flights that last less than a second and that move a bee from one point to another on the walls, floor, or ceiling of the cavity. A geometric pattern in the movements of scout bees is that early in the discovery inspection a scout walks primarily near the entrance during her journeys inside, whereas later she penetrates to the deepest recesses of the cavity (fig. 3.13). Three-dimensional reconstructions of the walking paths of individual scouts reveal that when the inspection is finished, a scout has walked 60 meters (200 feet) or more around the inside of the cavity and has covered all its inner surfaces.

I spent four weeks on Appledore Island in 1975, departing without solving the mystery of how scout bees evaluate candidate nest sites. Nevertheless, I felt

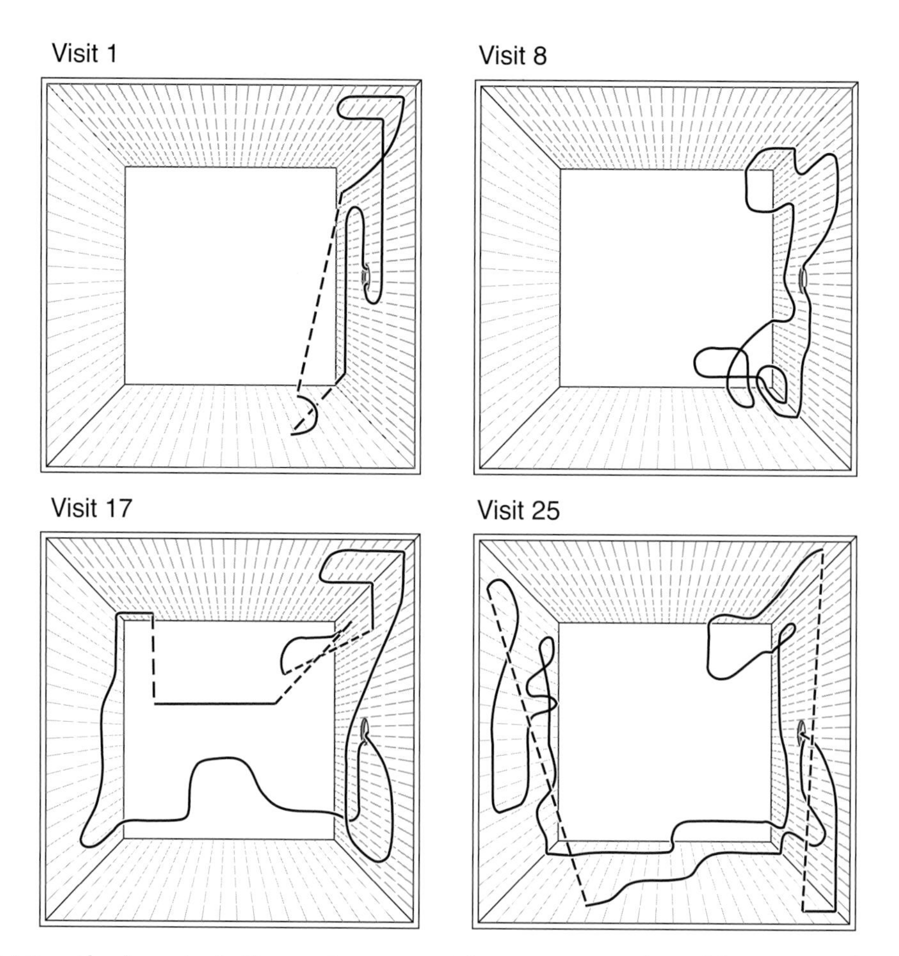

Fig. 3.13 Scout bee's method of inspecting a potential nest cavity is indicated by tracings of what a single scout did on four out of the 25 journeys inside that she made during her initial inspection of the observation nest box. Solid lines denote where the bee was walking; broken lines denote where she was flying.

I'd made a satisfactory progress. I had learned how to work with swarms on this offshore island, where stiff winds and persistent fog sometimes hampered my bee work, but also where the salty air, waves breaking over the rocky shores, and screaming gulls were always invigorating. And I had discovered how a scout bee behaves when inspecting a prospective homesite. Knowing this behavior would prove invaluable in planning future experimental work on Appledore. When I returned to the island in July 1976, I focused on the puzzle of how scout bees can measure the size of potential nest cavities, which are immense compared to the size of a bee. Cavity volume is the nest-site property that is perhaps most critical to a colony's long-term survival, since any colony occupying a hollow 10 liters or smaller cannot store sufficient honey to get through winter, so I suspected the bees have evolved a way to measure cavity volume accurately.

How do scout bees measure the volume of a cavity? Their extensive walking in the course of an initial inspection could provide the basis for an estimate, but another hypothesis was that they simply go inside and look around. I first performed experiments with nest boxes in which I could vary the interior light level (by changing the amount of light coming in through the entrance hole) and the traversable surface area (by coating the inner surfaces with Fluon, a teflon-type material that creates a waxy surface that bees cannot walk up) (fig. 3.14). I found that in order to measure a cavity's volume, scout bees need either interior illumination greater than 0.5 lux (about the illumination provided by a full moon) or inner surfaces that can be traversed freely. What are the conditions inside a typical tree cavity? Certainly the wood walls inside a cavity are easily traversed by a scout bee. To measure the light level in cavities of the sort inspected by scout bees, I built a model based on the measurements I had made of natural nests. It had a series of openings into which I inserted a light meter. I found the illumination to be less than 0.5 lux except near the entrance opening where some sunlight streams in. Evidently, in nature scout bees rely primarily on walking about in a prospective nest site to measure its volume.

To test this hypothesis more directly, I tried altering a scout's perception of a cavity's volume by manipulating the amount of walking required to move from point to point inside a cavity. To do so, I invented a bee treadmill, a cylindrical nest box mounted vertically on a turntable that enabled me to rotate the box smoothly while a scout bee was inside (fig. 3.14). By means of a window at the top, I could look inside and see which way the bee was walking; then I could turn the walls according to whether I wanted to increase or decrease the amount of walking required for her to complete a horizontal circuit. So, if a scout came in through the entrance and I rotated the walls in the direction she walked, she was carried along and quickly found herself back near the entrance. But if I turned

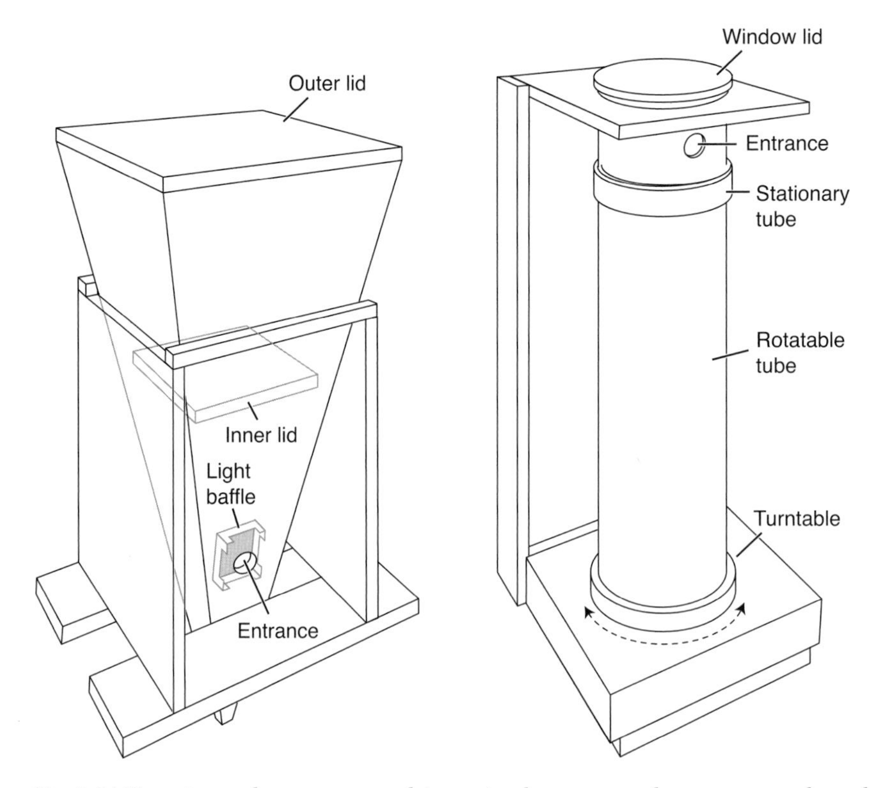

Fig. 3.14 Experimental apparatus used in testing how a scout bee measures the volume of a cavity. Left: Apparatus in which the cavity volume could be varied between 5 liters (with the inner lid down) and 25 liters (inner lid up). The light baffle made it possible to vary the amount of light coming in through the entrance hole, to see whether scouts could measure the size of the cavity without relying mainly on vision. Coating the interior wall surfaces made it possible to vary the amount of traversable surface area inside the box, to assess the importance of walking. Right: Apparatus in which the wall of the cylindrical cavity could be rotated to increase or decrease the amount of walking a scout bee had to do to circumnavigate the space.

the walls against her, she would need to walk much longer to make her way back around to the entrance opening. The entire device was mounted in my lightproof hut with a short tunnel connecting the entrance in the nest box to an opening in the hut's wall. The only light entering the nest box came in through its entrance, and it is likely that this bright spot provided each scout bee inside the box with a visual reference point, both for finding her way out of the box and for monitoring her progress in circumnavigating it.

The volume of this experimental box was 14 liters, on the boundary between an unacceptably small cavity and a suitably large one. If walking contributes to the perception of volume, then the first scout to discover the box should find it either more or less attractive than its true volume would merit, according to whether she had been made to walk more or less than she would in a normal 14-liter cavity. The assay of her evaluation of the box was the number of other scouts she recruited to visit the box; she should recruit more scouts if she found the box suitably large than she would if the box had seemed unacceptably small. That is precisely what I observed in four trials of this experiment: seven or nine recruits in 90 minutes when the bee walked lots, but only zero or one recruit in 90 minutes when the bee walked little. Evidently, only the bees that had taken "long walks" judged the box sufficiently roomy and recommended it to their fellow scouts with enthusiasm. It seems clear, therefore, that a scout's estimate of the volume of a cavity is proportional to the amount of walking she must do to circumnavigate it. Every step is a measurement.

Nigel R. Franks and Anna Dornhaus, biologists at the University of Bristol in England and the University of Arizona in the United States, have recently suggested a simple method by which a scout bee might judge a cavity's roominess using the information gained from the walking and flying movements that she makes inside the cavity. They point out that physicists have long known that for any open space the mean free path length (MFPL) of wall-to-wall lines drawn in all directions across the space is equal to four times the volume (V) of the space divided by its internal surface area (A): MFPL = 4V/A. Thus, volume is proportional to mean free path length multiplied by internal surface area: V = (MFPL x A)/4. It is possible that the extensive walks made by scout bees are a means of estimating a cavity's internal surface area. It is also possible that the short hopping flights made by scout bees—which I had reported but had not linked to the volume estimation process—are a means of seeing how far they can fly before hitting a wall, that is, a means of estimating the mean free path length. If both possibilities prove correct, then it may be that all a scout bee needs to do to be certain that a cavity is sufficiently spacious is to ascertain that the cavity provides a suitable combination of internal surface area and mean free path length. Certainly the results of my experiment with the rotating-wall nest box are consistent with this hypothesis;

forcing a bee to walk farther to return to the entrance (thereby increasing the bee's estimate of internal surface area?) resulted in a larger estimate of the box's volume. Franks and Dornhaus have proposed an ingenious experimental test of their idea, one that involves hanging a rigid curtain across much of the interior of a nest box and coating the curtain with Fluon so that scout bees cannot walk on it. This curtain will shorten the mean free path length of flights inside the nest box but will not change either its volume or its walkable surface area. One can then see if the bees behave as if the box has been shrunk. I hope the test will be performed soon, and that it will provide support for the proposed rule of thumb, for I think it's an elegant solution to a tough problem.

4
SCOUT BEES' DEBATE

The experience of democracy is like the experience of life itself—
always changing, infinite in its variety, sometimes turbulent
and all the more valuable for having been tested by adversity.
—*Jimmy Carter,* Address to the Parliament of India, *1978*

When a honeybee swarm chooses its future home, it practices the form of democracy known as direct democracy, in which the individuals within a community who choose to participate in its decision making do so personally rather than through representatives. The collective decision making of a bee swarm therefore resembles a New England town meeting in which the registered voters who are interested in local affairs meet in face-to-face assemblies, usually once a year, to debate issues of home rule and to vote on them, rendering binding decisions for their community. Of course, there are differences in how direct democracy works between bee swarms and town meetings. For example, the scouts in a bee swarm have common interests (e.g., all want to choose the best available homesite) and they reach decisions by building a consensus. The people in a town meeting, however, often have conflicting interests (e.g., some do and some don't want to help fund the town library), and they reach decisions by using the majority voting rule: each individual has one vote, all votes have equal weight, and the option that gets the majority wins. Another basic difference between bee swarm and town meeting is that a scout bee in a swarm, unlike a citizen in a meeting, cannot monitor each exchange within the group's debate and thereby have a synoptic overview of the discussion. Instead, a bee can only observe and react to the actions of her immediate neighbors in the swarm cluster, hence she operates without global knowledge of the information that percolates among her fellow swarm bees.

While these differences—common versus conflicting interests and local versus global knowledge—between bee swarm and town meeting are real, they do not overshadow several extremely important similarities between the direct democracies of honeybees and human beings. Firstly, in both the insectan and the human forms of this collective decision making, each decision about a future course of action reflects the contributions, freely given and equally weighted, of several hundred individuals. In other words, the control of the group's actions is distributed among many of its members rather than concentrated in a few leaders. Secondly, because hundreds of individuals are full participants, the group can acquire and process information from multiple sources simultaneously, even ones that are widely scattered. Consider, for example, the first stage of every decision-making process, where the critical challenge is to identify the available options. By virtue of having numerous individuals examining a problem and presenting possible solutions to it, both a bee swarm and a town meeting are much more capable than any solitary bee or single person in coming up with a broad range of alternative options. And the broader this range of options, the more likely it will include the one best option. Thirdly and most intriguingly, in both bee swarms and town meetings, the way the group selects its course of future action is by staging an open competition among the proposed alternatives. An individual proposes a possible way forward, each listener makes an independent assessment of the proposal and decides whether to reject or accept it, and those that accept it may announce their own support for the proposal. These endorsements often recruit still more supporters for this option. The better the proposal, the more supporters it will attract, and the more likely it is to gain sufficiently broad support to become the community's choice.

In the case of the house-hunting bees, the competition among the supporters of different proposals is often fierce, with some scout bees vigorously championing one lovely tree cavity while other scouts, just a few bees away on the surface of the swarm cluster, are enthusiastically advertising a second, third, or even fourth desirable dwelling place. (Later in the chapter, we will see which bees take up the "profession" of nest-site scout and what prompts these bees to step into this dangerous job. In a nutshell, nest-site scouts are elderly bees who were working as common foragers but then quit this line of work when they sensed that their colony, preparing to swarm, no longer needed additional food.) But it is always a "friendly" competition; the scout bees agree on what makes an ideal homesite, they are united in the goal of choosing the best available site, they share their information with full honesty, and ultimately they reach a complete agreement about their new residence. One valuable lesson that we can learn from the bees is that holding an open and fair competition of ideas is a smart solution to the problem of making a decision based on a pool of information dispersed across a group of individuals.

Lindauer's Swarms

The discovery that swarm bees use debates to aggregate dispersed information was made by Martin Lindauer in 1951 and 1952, when he received permission from Karl von Frisch to use all the swarms that emerged from the hives of bees kept in the garden behind the Zoological Institute at the University of Munich. Thus, at last, Lindauer was able to make a close study of what had aroused his curiosity back in the spring of 1949: dirty dancing bees on the surface of a swarm (fig. 4.1). The institute's bee colonies did their part, providing Lindauer with 17 swarms spread across the months of May, June, and July. Many of the swarms flew across the street to the Botanical Garden where they settled on various objects, and Lindauer charmingly named each swarm for its bivouac site: the Linden swarm on May 18, 1951, the Hawthorn swarm on July 9, 1951, the Balcony swarm on June 22, 1952, and so forth. His immediate aim was to determine whether the bees dancing on a swarm are nest-site scouts, but his ultimate goal was to understand how a swarm finds its new homesite. Knowing the value of starting an investigation of animal behavior with a "watching and wondering" phase, during which one patiently makes broad observations that can yield un-

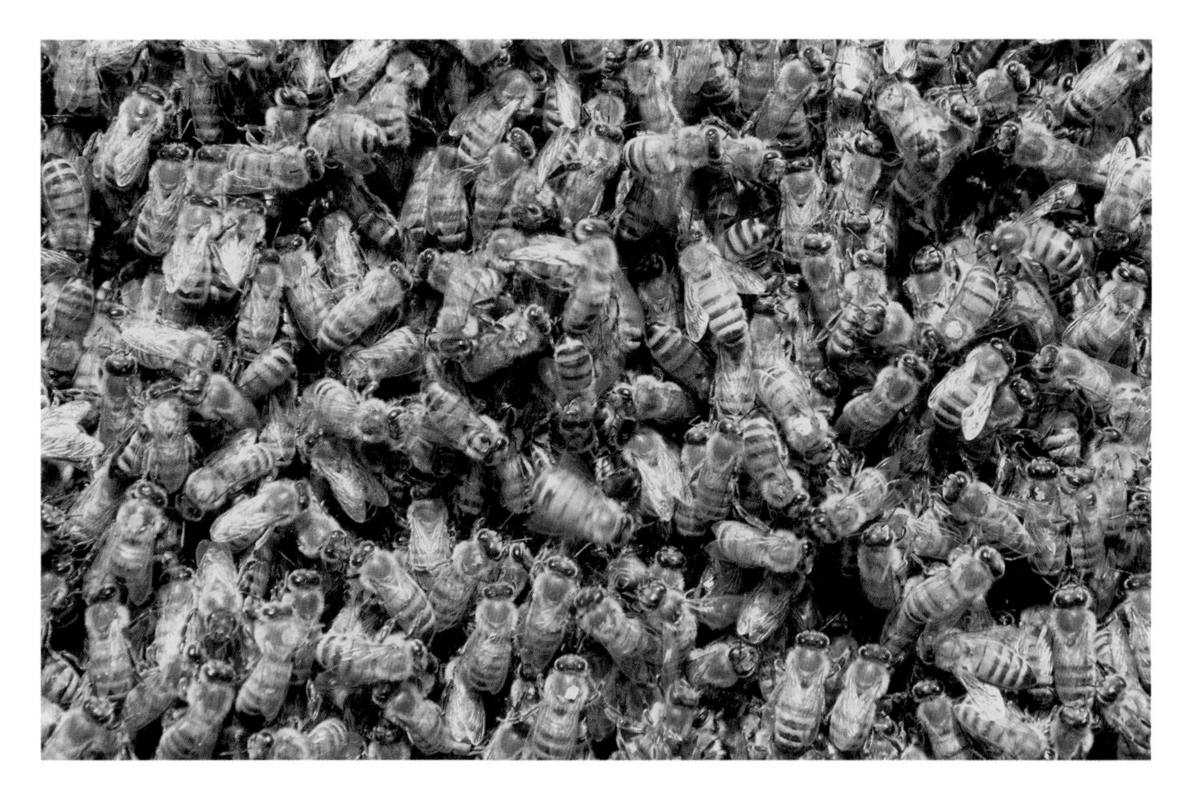

Fig. 4.1 Scout bee performing a waggle dance on the surface of a swarm.

expected discoveries, Lindauer watched from dawn to dusk the bees performing waggle dances on the surface of each clustered swarm. Knowing too that he wanted his dancing bees to be identifiable individuals, so that he could begin to enter the world of a bee in a swarm, he applied colored paint dots to the back of each dancing bee. Karl von Frisch had devised a marking scheme back in the 1910s that enables one to number bees from 1 to 599 with just 5 paint colors, and Lindauer used this system, deftly daubing with a tiny paint brush one to four tiny dots of shellac paint on the thorax of each dancing bee.

Lindauer's endeavor to record the dances on a swarm started easily enough, for few bees danced and their dance performances were intermittent. So at first Lindauer could record in his notebook the time of each bout of dancing, the identity of the dancer, and what location this individual advertised with her dance.

In time, however, his task of recording the swarm bees' dance activity became almost impossibly difficult; sooner or later he faced a dozen or more bees dancing simultaneously on the swarm. He coped with this overwhelming display of dancing bees by becoming more selective in what he recorded; he noted only the time he saw each new (not yet labeled) dancing bee and the location indicated by each new bee's dance. It was exhausting to single-handedly observe, label, and record the dancing swarm bees for hour after hour, sometimes for days on end. But it was immensely revealing. In chapter 1, we saw how Lindauer tested his hunch that the bees dancing on swarms are nest-site scouts advertising potential nesting cavities. He found that the site indicated unanimously by the dancing bees shortly before a swarm flew to its new homesite matched the address of their new dwelling place. Even more marvelous was Lindauer's discovery that a swarm's scouts conduct a vigorous debate to select their new home.

Figure 4.2 shows an example of one of these debates. It was held on the Eck swarm, a swarm that left its parental hive at 1:35 in the afternoon on June 26, 1951, soon settled in a privet bush, and then hung out there for nearly four days while its scout bees went about choosing a dwelling place. On the first day, Lindauer observed and labeled just two dancing scout bees, between 1:35 and 3:00 p.m. One bee reported a candidate homesite approximately 1,500 meters to the north, while the other proposed a second site 300 meters to the southeast. By 3:00 p.m. the sky had filled with dark rain clouds and the air had became cooler, so the scout bees ceased their explorations for the day. The next day the scouts remained inactive until late in the morning, when the clouds parted and bright sunshine returned. Figure 4.2 shows that Lindauer labeled 11 new dancers between noon and 5:00 p.m., and that three of them advertised the site 1,500 meters to the north, two others advertised the site 300 meters to the southeast, and the remaining six indicated six other sites in various directions and distances. Clearly, no agreement among the dancing bees was reached during this second day of debate. On the third day, the weather was mostly rainy and Lindauer recorded just two new dancers, both in late morning. One advocated the site to the north, thereby strengthening this site's slim lead among the dancers (with five bees total, so far), and the other reported a new site, about 400 meters to the southwest.

day, June 13, that this symmetry began to break. For some reason, the strength of dancing for the southwest site must have weakened slightly, so fewer new dancers for this site were marked, and over the afternoon the mustering of new dancers for the northeast site surged ahead of that for the southwest site: first 25 versus 9 new dancers (12:00 to 2:00 p.m.), then 41 versus 7 new dancers (2:00 to 4:00 p.m.), and eventually 34 versus 0 new dancers (4:00 to 5:00 p.m.). We see that by the end of the third day an agreement had been reached, but, alas, it was too late in the day for the swarm to undertake its flight to its new home. (Swarm's rarely take flight after 5:00 p.m., perhaps to avoid risking a move without a long period of daylight ahead. A swarm's queen may need to make an emergency landing to rest, and when this happens it can take the workers in an airborne swarm over an hour to halt their group flight, locate their missing queen, and reassemble around her.) The Propyläen swarm's delayed decision-making forced it to stay put until the next day, which turned out to be cool and rainy, so it was not until the afternoon of June 15, fully four days after the swarm left its parental hive, that the bees finally flew off to their new residence in the northeast.

Lindauer even observed one swarm that failed to reach an agreement, that is, there never arose a dancer group that so dominated the deliberations that Lindauer marked new dancers for only one site. It was the Balcony swarm of June 22, 1952 (fig. 4.4). As in the Propyläen swarm, its scout bees got into a balanced competition, with one group of dancers advocating a site 600 meters to the northwest and a second group favoring instead a site 800 meters to the southwest. For four hours (12:00 to 4:00 p.m.) neither group managed to gain a decisive lead. Nevertheless, at 4:10 p.m. the swarm lifted off and then did something that Lindauer could scarcely believe even though he was seeing it with his own eyes. In his words, "The swarm . . . sought to divide itself. The one half wanted to fly to the northwest, the other to the [southwest]. Apparently, each group of scouting bees wanted to abduct the swarm to the nesting place of its own choice." And each group partly succeeded, for half of the airborne bees started moving southwest toward the main railway station while the other half began trending northwest toward the Karlstrasse (see fig. 1.6). But neither group managed to keep going in its desired direction, perhaps because each lacked the queen, and eventually the two clouds of swirling bees reunited in the air where

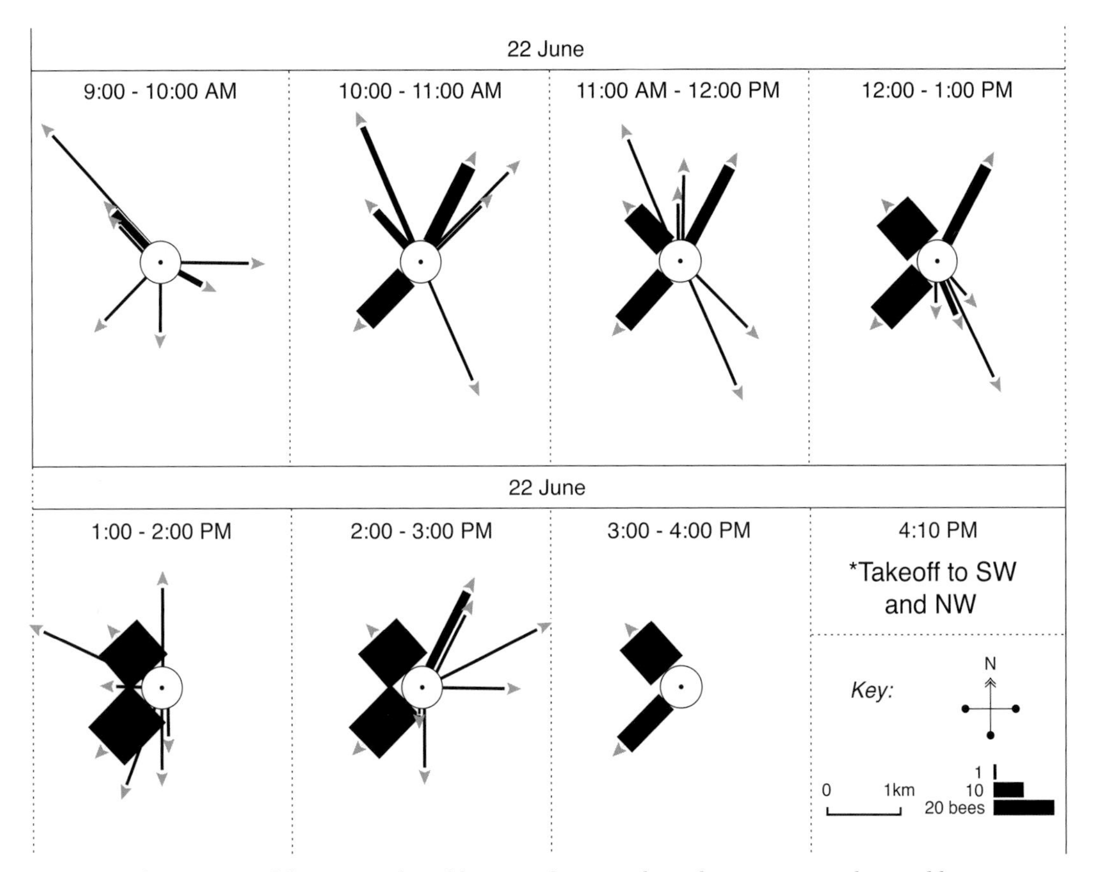

Fig. 4.4 The pattern of dances produced by scout bees on the Balcony swarm, observed by Lindauer in June 1952. This swarm's scouts never reached an agreement.

they had started. Then a remarkable tug-of-war between the two groups began. Over the next half hour, one group tried again to advance to the northwest, going off 100 meters before coming back, then the other group pushed 150 meters to the southwest, but then it too returned to the original site. At this point the bees resettled on the balcony where they had previously been clustered. Sadly, the swarm's queen had become lost during the aerial tug-of-war, and over the next several hours Lindauer watched the swarm cluster gradually dissolve as the queenless bees drifted home to their mother hive.

The history of the ill-fated Balcony swarm highlights several features of honeybee swarms and their decision-making process. From the tragedy of the queen's loss and the swarm's dissolution, we are reminded that the survival of a swarm's queen, who carries the new colony's genes in her ovaries and her spermatheca, is absolutely critical to a swarm's success. From the failure of the scouts to reach an agreement, we are shown how swarms are not infallible in their decision making. Occasionally, a swarm will produce a split decision, though usually this is just a temporary situation and the swarm manages to resolve the problem. We will see in chapter 7 that when a split decision arises, a swarm will normally take off, fail to move to either site, resettle, and then conduct further debate leading to an agreement. Lindauer watched 17 swarms of which just two (the Balcony and Moosacher swarms) produced split decisions. Only the Balcony swarm never reached agreement, and this was because it lost its queen, so it seems clear that complete failure in swarm decision making is rare. Finally, from the launch of the Balcony swarm into flight without prior consensus among the dancers, we are given a strong indication that dancer consensus, though conspicuous to the human observer, is not what the bees themselves monitor to know when they should switch from *making* a decision to *implementing* a decision. How the bees actually make this switch will be revealed in chapter 7.

My Swarms

A great scientific discovery is one that gives rise to shining insights whose light dispels obscurities, opens up new paths, and reveals unknown horizons. Martin Lindauer made such a discovery when he found that a honeybee swarm chooses its future home through a debate in which the nest-site scouts express their arguments in waggle dances. He elucidated the function of dances on swarms, he blazed a trail toward understanding the swarm bees' system of decision making, and most importantly he guided us into a whole new scientific territory: sophisticated group decision making by nonhuman animals.

Lindauer was certainly a pioneer in behavioral biology, and like all pioneers, the time and tools available to him were insufficient to explore fully the new terrain he had discovered. It is not surprising, therefore, that his investigation of honeybee democracy was incomplete in many ways. We can see this perhaps most clearly in the way he was limited by his equipment (notebook, watch, and paint set) to recording only the appearance of new dancers on his swarms. Ideally, he would have recorded all the dancers (new and old) that appeared on his swarms during each stage of the decision-making process, so that he could have acquired a full picture of the dynamics in the dancing for the alternative sites. This would have shown how the total number of supporters for each proposed site—not just the appearance of new supporters—changed over time, and how in the end presumably just one site was being advocated by the dancing bees. The records shown in figures 4.2 and 4.3 indicate that shortly before swarms flew away the winning site was being advertised by all the new dancers, but these records don't tell us whether in the end the winning site was being advertised by *all the dancers*, that is, all the old ones and all the new ones. Does the decision-making process in fact end cleanly, with essentially every dancer supporting a single site, the winning site? Lindauer suggested that it does; he wrote that the scouts for the losing sites ultimately "gave up their recruitment," presumably by ceasing to dance, but he did not show that they stopped dancing. Also, he did not show when the advocates of the losing sites stopped dancing or how they stopped dancing. It would be, therefore, very informative to have a complete picture of the dancing on a swarm. It would also be extremely desirable to have complete records of the behaviors of individual dancers, so that one could track each dancer's actions following her first bout of dancing. Does she perform multiple bouts of dancing? Is the total amount of her dancing related to the quality of the site that she is advertising? And if she stops dancing, how does she decide to do so? Does she quit on her own regardless of what is happening around her, or only after encountering another bee performing a more vigorous dance? Lindauer's study is an amazing first reconnaissance of how house-hunting honeybees perform their collective decision making, but it left unanswered countless questions about the rules of procedure that nest-site scouts follow when conducting their deliberations.

In 1996, I decided to tackle these questions. This was nearly 20 years after I had finished my PhD thesis research on the nest-site preferences of honeybees and the way they estimate the volume of a prospective nest cavity. Why hadn't I started addressing the gaps in Lindauer's work back in the mid 1970s? It was because I didn't see how I could get the video recording equipment that I knew I'd need to go beyond Lindauer's analysis of swarm decision making. In those days, a color video camera, recorder, and monitor (they were all separate units back then) cost many, many thousands of dollars, a price that far exceeded the budgets of the small grants I could get as a beginner scientist. So I changed the focus of my research, but without abandoning the topic of how social animals make collective decisions. Rather, I switched to studying another form of collective decision making by the bees: how a honeybee colony wisely deploys its foragers among the kaleidoscopic array of flower patches in the surrounding countryside. This is a different sort of collective choice, for whereas a homeless swarm makes a "consensus decision" about which *single option* (candidate nest site) it will choose, a foraging colony makes a "combined decision" about how to allocate its foragers among *multiple options* (candidate food sources). I was attracted to the puzzle of colony decision making about forager allocation partly because it looked fundamentally similar to nest-site choice—being based on competition among groups of dancing bees advertising different options (food sources rather than nest sites)—and partly because it looked more tractable than nest-site choice. Swarming is an ephemeral phenomenon that lasts at most a few days, whereas foraging goes on all summer long. So for about 15 years I explored with pleasure how the bees in a hive work together as a unified whole in gathering their food, especially how they wisely distribute themselves among flower patches. In 1995 I summarized this body of research in a book, *The Wisdom of the Hive*, and with this behind me I looked forward to resuming my exploration of collective decision making in honeybee swarms.

The starting point was clear. I should obtain a complete record of the scout bees' dances throughout a swarm's choice of its new home in order to get a full picture of how a debate among scout bees unfolds. This broad description of the scout bees' behavior at the swarm would fill the gaps in the description provided by Lindauer, and probably would also yield important discoveries, as indeed it did. Unlike Lindauer 40 years before, or myself 20 years before, I now had sophisticated video recording and slow-motion playback equipment that made it possible to build a comprehensive record of the scout bees' dance activity. Also, I now knew how to label thousands of bees for individual identification with plastic color-number tags glued on the thorax and paint marks placed on the abdomen (fig. 4.5), a skill that I had honed in my studies of honeybee foraging. With each

Fig. 4.5 Worker bees labeled for individual identification.

bee in a swarm so labeled, I figured that it should be possible to trace each individual's history of dancing throughout the decision-making process. Success in this project would require, however, a formidable amount of painstaking labor. Thousands of bees had to be individually labeled for each swarm, the swarm and video equipment had to be tended throughout each swarm's choice of a home, and the information about each dance (the dancer's ID, the site being advertised, and the dance's duration) had to be extracted manually from the video recordings. It was my immense good fortune to be joined in this endeavor by Susannah Buhrman, an extremely bright and indefatigable undergraduate student at Cornell. She proved an indispensable partner in this project. Working together throughout the summer of 1997, we achieved success.

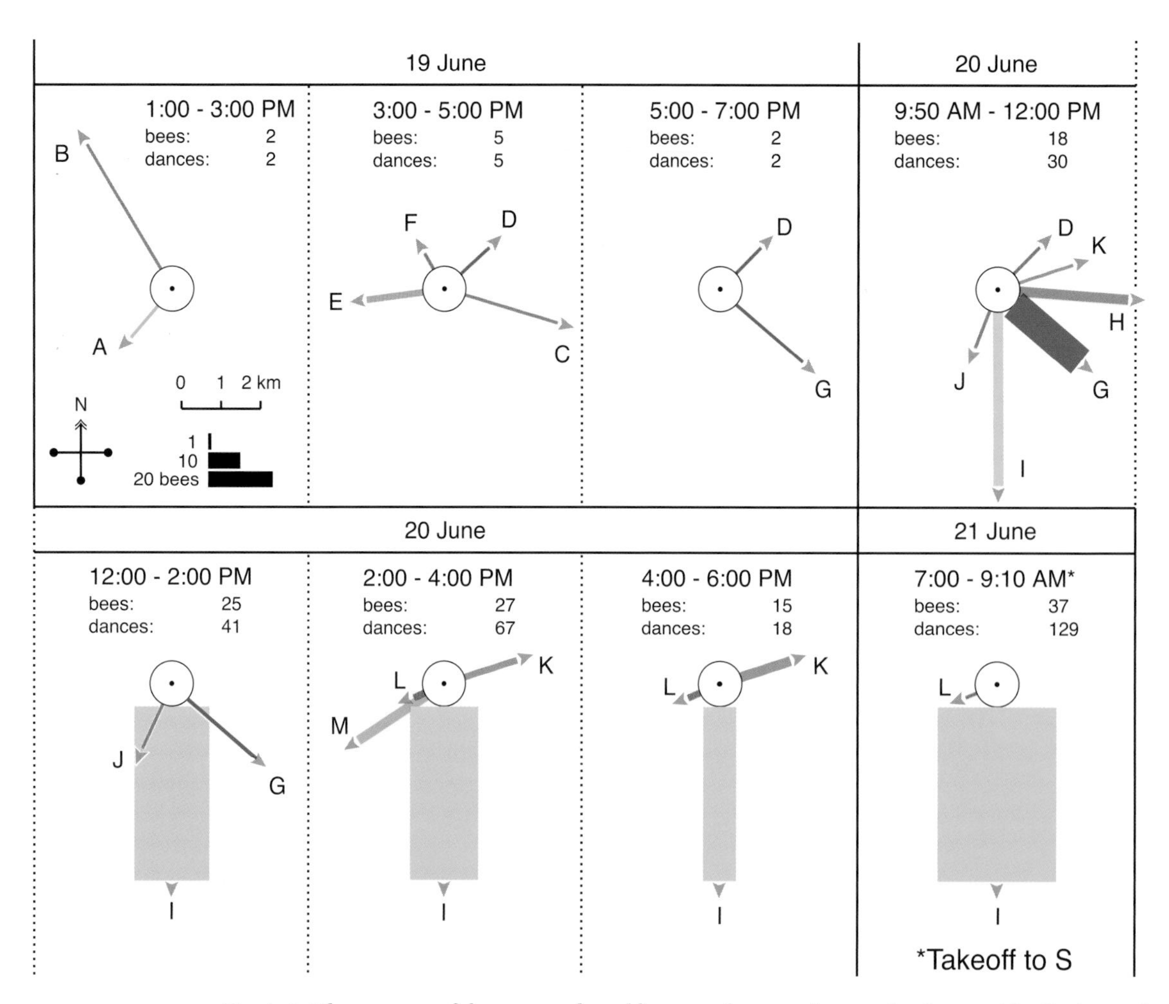

Fig. 4.6 The pattern of dances produced by scout bees on Swarm 1, observed by Seeley and Buhrman in June 1997. The width of each arrow shows the total number of dancers for the site during the time period indicated. This swarm's scout bees quickly reached a consensus.

Susannah and I eavesdropped on the deliberations among the scout bees on three swarms, and we obtained a complete record of the dances performed for each swarm's decision making. Figure 4.6 shows the debate recorded on Swarm 1, which was set up at 10:00 a.m. on June 19. We see that nest-site scouts started reporting discoveries between 1:00 and 3:00 p.m. and that by the end of the day seven candidate nest sites (A–G) had been raised for consideration, though none of these had elicited enthusiastic support. The next day the scout bees were livelier. By midday, four additional sites (H–K) had been entered into the discussion and three sites—G (2,200 meters to the southeast), H (2,600 meters to the east), and I (4,200 meters to the south)—had received endorsements by multiple dancing bees. Site G appeared to be developing a lead, for nine bees had advertised it, but no site was yet dominant in the dancing. The situation changed markedly between 12:00 and 2:00 p.m. Now site I rose to prominence, supported by 23 out of the 25 dancing bees. This situation persisted for the remainder of the afternoon, though still two more possibilities (sites L and M) were presented and the dancing bees showed support for sites K, L, and M as well as site I until the end of the day. The next morning, however, there was a clear consensus among the dancing bees in favor of site I, and at 9:10 a.m. the swarm took off and flew to the south, no doubt with site I its destination.

The debate on Swarm 1 proceeded in a manner reminiscent of what Lindauer described for his Eck swarm. During the first half of the decision-making process, the scouts reported numerous candidate sites located at various directions and distances from the swarm cluster. Then, in the second half of the debate, the dances of the scout bees quickly and smoothly became focused on just one site. Ultimately, there was virtual unanimity among the dancing bees, and the swarm moved to the agreed-upon site. It is worth stressing that figure 4.6 depicts, for each time interval, the number of *total* dancers for each site, not just the number of *new* dancers for each site. We can be confident, therefore, that the conclusion of this swarm's decision making was marked by a real consensus among the dancing bees.

The most interesting scout bee debate that Susannah and I observed occurred in our Swarm 3; in this one there was a strong competition between two groups of dancers, and for many hours it was unclear which faction would emerge victorious (fig. 4.7). This swarm was set up at 2:30 p.m. on July 19, but it was not until the middle of the next day that scout bees started advertising potential nest sites with their dances. Six sites (A–F) were announced between 11:00 a.m. and 1:00 p.m., one of which (site A, 2,200 meters to the east) quickly developed a strong lead with eight dancing bees promoting it. Over the next four hours three more

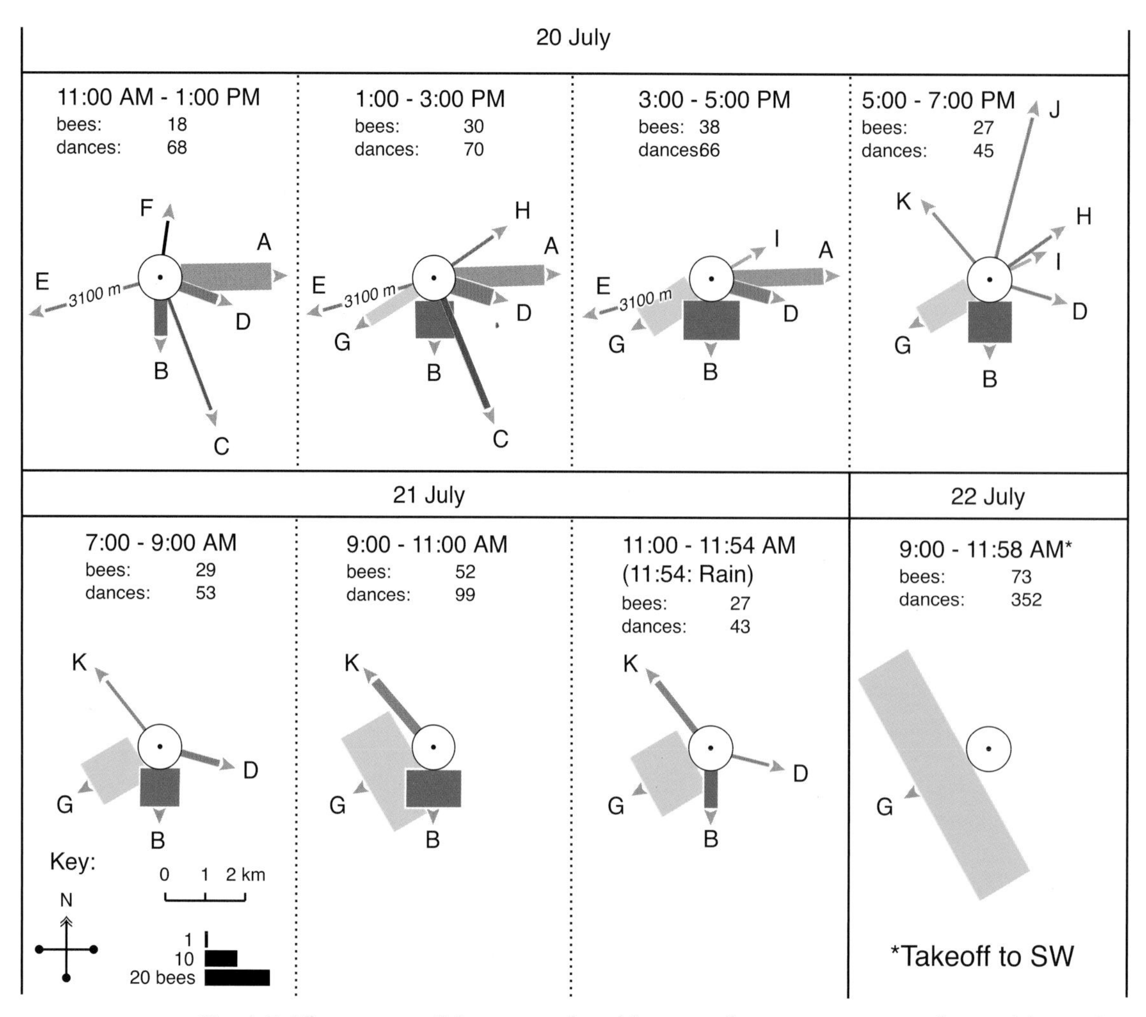

Fig. 4.7 The pattern of dances produced by scout bees on Swarm 3, observed by Seeley and Buhrman in July 1997. This swarm's scout bees reached a consensus, but only after a prolonged competition between the scouts from sites B and G.

possible home sites (G, H, I) were entered into the discussion, and four sites (A, B, D, G) received strong consideration, each being advertised by several bees. Site A, though, was losing its initial lead as sites B (900 meters to the south) and site G (1,400 meters to the southwest) gained more and more support. Between 3:00 and 5:00 p.m., only four bees danced for site A, whereas 17 bees danced for site B and 10 did so for site G. At 5:00 p.m. it seemed that the contest among dancer groups on this swarm was still wide open. This situation changed dramatically during the remaining two hours of dance debate that day, for although bees performed dances for seven sites during these last two hours, including two new sites (J and K), only sites B and G received the support of multiple dancing bees. Susannah and I could see that the supporters of these two sites had managed to gain wide leads over the bees promoting the other nine sites, and we made bets on whether the B group or the G group would prevail the next day. I bet on site B and Susannah on site G. Whoever would win the bet would be treated to a triple scoop ice cream cone at the new Ben and Jerry's in Ithaca.

The tension was high the next morning. We arrived at the lab shortly after sunrise, got our recording equipment set up before the scout bees could resume their debate, and waited eagerly to see who would win our little bet. For the first two hours, 7:00 to 9:00 a.m., we both remained hopeful, because both sites were being advertised by about a dozen dancing bees. Starting around 9:00 a.m., though, my optimism began to ebb, for the dancers supporting Susannah's site G began to build a commanding lead over those for my site B, with 32 versus 17 bees respectively between 9:00 and 11:00 a.m., and 20 versus 4 between 11:00 and 11:54 (when it started raining). Somehow the site G bees had managed to overwhelm those for site B. The rain continued throughout the afternoon, during the night, and until about 8:00 a.m. the next day. The scout bees resumed dancing a little after 9:00 a.m., and now they showed unanimous support—73 out of 73 bees!—for site G in the southwest. Shortly before noon the bees flew off to the southwest, and shortly after noon Susannah and I motored off to Ben and Jerry's.

It was a great pleasure to watch the dance competitions among scout bees, but it was an even deeper pleasure to analyze the diagrams like figures 4.6 and 4.7 that we prepared many weeks later, after we had extracted all the information we needed from our 48 hours of video recordings. These diagrams gave us a crystal clear picture of the main features of the scout bees' decision-making process. First, they showed that the bees' debates tend to start slowly with an information accumulation phase during which scout bees put a sizable number of widely scattered alternatives "on the table" for discussion. In the three swarms that Susannah and I watched, the number of sites considered was 13, 5, and 11. These sites were located in various directions and at various distances (200 to 4,800 meters) from the swarm cluster, which indicates that the intrepid scouts from these swarms had searched some 70 square kilometers (about 30 square miles) of countryside for possible dwelling places. Most of the candidate sites were introduced during the first half of the deliberations, but as we can see with sites L and M in Swarm 1 (fig. 4.6), sometimes a few got introduced rather late in the discussion. Certainly a swarm does not manage to identify all its alternative options simultaneously, but as we will see in chapter 5, this asynchrony usually does not lead to swarms making poor decisions.

Second, the plots of the dance records showed that the scout bees' debates end with all or nearly all of the dancing bees advocating just one site, that is, showing a consensus. A burning question is, therefore, how does the fierce competition among groups of bees favoring different options get transformed into a harmonious agreement? Specifically, how is it that the number of dancers builds up for one site (generally the best one, as we will see in chapter 5) while at the same time it falls to zero for all the other sites? We will see how the bees accomplish these things, using some nifty tricks, in chapter 6.

Third, our analysis showed that the bees' decision-making process is a highly distributed and thus a democratic one, involving dozens or hundreds of individuals. Susannah and I observed 73, 47, and 149 bees performing dances in the three swarms we studied. These counts, however, probably underestimate the typical number of dancing bees in a swarm. This is because we used unusually small swarms—with only 3,252, 2,357, and 3,649 bees—to keep doable our task of individually labeling the bees. Swarms in nature generally contain 6,000 to 14,000 bees. The mean percentage of dancing bees in our labeled-bee swarms was 2.8 percent, which is similar to the 5.4 percent figure for natural swarms reported by David Gilley, another dedicated Cornell undergraduate student who investigated the mystery of the identity of scout bees (next section). Given that 3 to 5 percent of the bees in a swarm participate in the dance debate, we can estimate that a typical swarm of some 10,000 bees will have approximately 300 to 500 individuals contributing to the decision-making process.

Intrepid Explorers

The profession of nest-site scout is performed only a few days each year, usually in the late spring or early summer. This fact, together with the fact that worker bees have short life spans—only three to five weeks during the warm months of the year—tells us that many generations of bees will pass without the need arising for individuals to explore for new accommodations. And yet when a colony prepares to cast a swarm, a small fraction of its workforce springs into action as nest-site scouts. These intrepid explorers are the prime movers of the whole swarming process. They determine when the swarm leaves its mother hive (as discussed in chapter 2), they make the swarm's life-or-death choice of a suitable nesting cavity, they trigger the swarm's takeoff to fly to its new home (see chapter 7), and they steer the swarm during its flight (see chapter 8). Who are these all-important bees? And what stirs them to action?

Evidently, scout bees are forager bees that have radically switched their behavior so that instead of seeking bright blossoms they search for dark crevices. The first evidence that nest-site scouts are reconfigured foragers came from an experiment conducted by Martin Lindauer. On May 11, 1954, Lindauer set up a colony in a locale east of Munich where flat fields stretch to the horizon and few trees and houses offer nesting cavities. There was, however, plentiful forage for the bees and within a week they began filling the combs in their hive with brood, pollen, and honey. Lindauer expected them to swarm shortly and eventually they did so, casting a swarm on May 27. Ten days before this, on May 17, Lindauer had set up a table 250 meters (820 feet) from the hive and on it he had placed a feeder filled with rich sugar syrup (granular sucrose dissolved in honey). In a few days, he had more than 100 bees from his hive foraging eagerly at his feeder, and he had each one labeled with paint marks for individual identification. Next, on May 22, he placed two artificial nest sites beside the feeding table: a straw skep and a wooden hive (fig. 4.8). Over the following few days, Lindauer started to see curious changes in the behavior of the bees visiting his feeder. First the eagerness of their foraging decreased. Fewer and fewer of his labeled bees made trips to the feeder, and those that kept coming visited less and less often. Sometimes they

Fig. 4.8 Lindauer's arrangement of two artificial nest sites beside a small table bearing a feeder filled with rich sugar syrup.

sucked only hesitantly at the rich sugar syrup. Eventually, on the morning of May 25, Lindauer noticed that his "foraging bees made only a pretense of coming to the feeding dish. They did sip very briefly at it, but then they flew up *and buzzed around in the near vicinity for some time.*" A knothole in a nearby oak tree drew their attention, as did the two artificial nest sites Lindauer had provided. Over the afternoon, six of his labeled foragers (Bee 73, Bee 100, Bee 106, Bee 113, Bee 119, and Bee 156) conducted 15 inspections of the skep and eight of the hive. There could be no doubt: some of his foragers had become scouts!

The second indication that nest-site scouts are converted foragers comes from a study conducted by Dave Gilley, the gifted undergraduate student who joined my laboratory and quickly fell in love with the bees. To earn an honors degree at Cornell, a student majoring in biology must write a senior thesis based on original research. Dave approached me in the spring of his junior year about attempting an honors thesis project with the bees. I suggested he probe further the mystery of who becomes a nest-site scout, and he happily accepted. Lindauer had shown that *some* scouts were previously foragers. Dave wanted to see if *all* or *most* scouts were previously foragers. If so, then the scouts should be among the oldest bees in a swarm, for it is well established that foragers are the oldest bees in a hive. Dave tested this prediction by setting up five small colonies of bees in early May 1996, and every three days, from May 5 to July 22, adding to each one a cohort of 100 bees that had just emerged from brood combs held in an incubator (0-day-old bees). All the bees in each age cohort were labeled with a paint dot of a particular color indicating the group to which they belonged. Over the next several weeks, as Dave filled the colonies with colorful bees, the bees filled their combs with brood, pollen, and honey. Then, one by one in June and July, the colonies swarmed. Once a swarm had settled into a cluster outside the laboratory building, Dave would watch it patiently for paint-marked bees performing dances, and each time he saw such a bee he would record her age and give her another paint mark (to avoid counting this bee again). Once he had sighted 50 or so nest-site scouts of known age, he collected the entire swarm, narcotized the bees with carbon dioxide before placing them in a freezer, and finally picked over the dead bees to count how many were in each age cohort. These counts enabled him to calculate the age distribution of the nest-site scouts that would be expected if they were drawn randomly from among all the known-age bees in the swarm. Figure 4.9 shows typical results from one swarm. We see that the nest-site scouts included many more older—that is, forager age—bees than would be expected if the known-age scouts had been drawn at random from the pool of known-age bees in the swarm. These findings support the idea that scouts come largely, if not entirely, from the ranks of a colony's foragers. Both scouts and foragers make long-distance excursions from a central location (swarm or hive) and then must find their way home, so it is easy to imagine that bees with foraging experience make the best scouts.

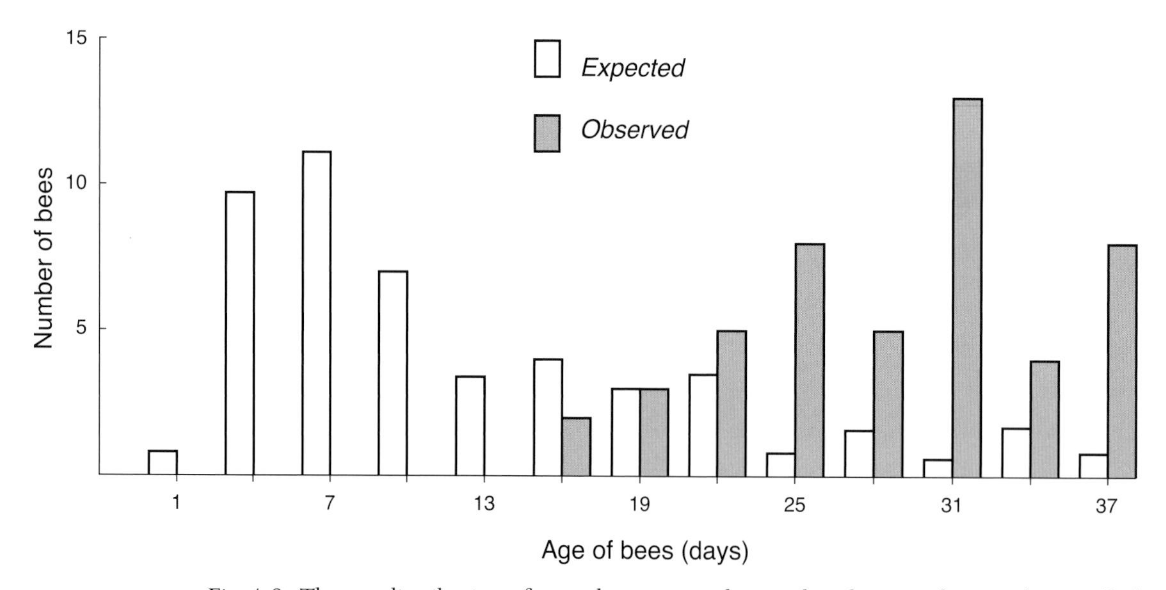

Fig. 4.9. The age distribution of scout bees versus the age distribution of swarm bees. Filled bars show the number of scouts observed in each age cohort. Open bars show the number of scouts expected in each age cohort if the scouts had been drawn randomly from the bees of the swarm.

Having foraging experience evidently prepares a bee for the special job of house hunting. This certainly cannot be the whole story, however, because many foragers never go searching for real estate opportunities. We now know that having certain genes also predisposes a bee to serve as a nest-site scout. Biologists have shown repeatedly, and in many animal species, that differences in behavior among individuals arise from differences in both their genes and their experiences, so it is not surprising that scouts and nonscouts in honeybee swarms differ in "nature" (genes) as well as "nurture" (experience). The need to have the right genetic stuff in order to become a scout bee was shown by two behavioral geneticists, Gene E. Robinson and Robert E. Page Jr., now professors at the University of Illinois (Urbana-Champaign) and Arizona State University. They established three colonies, each of which was headed by a queen that had been instrumentally inseminated with the semen of three unrelated drones (A, B, and C). The three sperm donors for each queen carried distinct genetic markers so that the investigators could determine which drone (A, B, or C) fathered any given worker in a colony. Robinson and Page then prepared artificial swarms (the method is explained below) from their colonies, set these swarms up outdoors, and collected about 40 scouts (dancers) and 40 nonscouts (nondancers) from each swarm. Finally, they conducted a paternity analysis of each collected bee and statistically analyzed their findings to see if the offspring of some drones were more likely to be scouts than were the offspring of other drones. In two of the three swarms they found that yes, the offspring of the three drones differed dramatically in the likelihood of becoming a nest-site scout. For example, in one swarm, one drone fathered over 60 percent of the scout bees even though he fathered less than 20 percent of the worker bees overall. One wonders what it was about this drone's genes that gave his daughters their proclivity to set out as enterprising house hunters, going boldly where no bee had gone before.

Of course, it is only when a colony is in swarming mode that some foragers, especially those endowed with genes fostering exploratory behavior, adopt the special role of house hunter. How do these bees know when it's the right time to change their occupation from forager to scout? One hint of how the bees might do this comes from what we humans must do to concoct an artificial swarm for an experimental study, such as the one just described by Gene Robinson and Rob Page. Basically, this is a matter of rendering a queen and a contingent of her workers *homeless but not hungry*. To do so, you first search through a hive of bees until you locate the queen and then you sequester her safely in a matchbox-size "queen cage." Next, using a large funnel, you shake several thousand worker bees off the hive's combs and into a shoebox-size "swarm cage" that has bottom, top, and ends made of wood, but sides made of window screen (for ventilation). At this point, you suspend the queen cage inside the swarm cage, so the worker bees have their queen, and you close the top of the swarm cage, so the bees are contained. Finally, you feed the caged bees lavishly by brushing sugar syrup onto the screen sides of the swarm cage. It is absolutely essential to feed the bees until they are sated and then to keep them full of food for several days. If you don't, when you shake the worker bees from the cage you will see that the workers cluster conveniently wherever you have mounted the queen (still confined in the queen cage), but you will not see scout bees springing into action. I know this from failures experienced firsthand. When I began making artificial swarms, I sometimes made the mistake of not feeding a swarm sufficiently before setting it up. Then I would sit beside the swarm for days waiting for the scout bees to start their dancing, wondering why they didn't. It seems that a critical stimulus for inducing a food collector to transform herself into a house hunter is that her stomach has been filled with food for a few days.

Lindauer observed this transformation from hungry forager to sated scout during the study described above that he performed in May 1954. The colony used in this study was housed in a glass-walled observation hive so Lindauer could watch the behavior of foragers both inside and outside the hive. When he established his sugar water feeder 250 meters from the hive on May 17, there was little natural forage available, and the bees that found his feeder danced vivaciously when they came home laden with sugar syrup. Over the next few days he labeled more than 100 foragers at the feeder. From May 22 on, however, the flowers of horse chestnut trees (*Aesculus hippocastanum*) provided plentiful nectar, the bees gradually filled their hive's combs with honey, and when Lindauer's foragers returned home they had difficulty finding hive bees who would accept their loads of sugar syrup. It is now well established that when returning foragers experience difficulty unloading their nectar, they lose their zest for dancing and foraging. In the extreme condition of a strong colony that has its combs filled with brood and food (hence it is primed to swarm), it is likely that foragers will find it impossible to unload their nectar and will linger at home with their stomachs bulging. This forced inactivity may stimulate a few foragers, those who are constitutionally inclined to explore, to turn to nest-site scouting. I find it extremely suggestive that Lindauer started seeing some of his labeled foragers exploring his nest sites, not exploiting his feeder, a few days after he started noticing most of his previously active foragers sitting around idly, either in some quiet spot inside the hive or in the "beard" of bees hanging outside the entrance. Anecdotal observations like these are the perfect springboard for an experimental investigation designed to test conclusively whether it is a persistently full stomach per se, or something else associated with forced indolence, that informs foragers to become scouts. Students take note.

5

AGREEMENT ON BEST SITE

Love quarrels oft
in pleasing concord end.
—John Milton, Samson Agonistes, *1671*

In the previous chapter we saw how the quarrels among scout bees, like those among human lovers, "oft in . . . concord end." Now we will see if the agreements reached by the bees are "pleasing." That is, when the dancing bees reach a consensus about their new homesite, are they apt to have chosen the best site? The answer is yes! But before looking at the evidence that a swarm usually chooses the best of the many candidate sites the scouts discover, let's first consider the structure of the choice problem faced by the house-hunting bees. This will sharpen our appreciation of a honeybee swarm as a democratic decision-making body.

A swarm of bees selecting a nest cavity faces a decision-making problem akin to what a person faces when choosing a place to live. This is a complex choice problem, for it is one where there are numerous alternative solutions (e.g., houses or apartments) and each one has many attributes (e.g., neighborhood, number of bedrooms, and such). And as is true for all decision-making problems, finding a good solution is a twofold process: first identify the possible alternatives, then choose among them. In an ideal world, the decision maker would be able to learn about all the alternatives and all the attributes of each, calculate the value of each alternative in light of all its attributes, and rationally choose the one with the highest value. Doing all these things will produce optimal decision making. In the real world, however, truly optimal decision making rarely happens because decision makers must pay costs in time, energy, and other resources to acquire and process information, and these costs usually preclude making the decision using all the relevant information. For example, someone hunting for an apartment in a large city would have to expend excessive time, money, and mental effort to survey the entire market of available rental properties, evaluate them all, and make the perfect choice.

Given that decision makers do not possess unlimited time, boundless resources, and infinite powers of reason, psychologists and economists now recognize that real world decision making—often called bounded rationality—relies on simplified mechanisms of choice, termed *heuristics*. These generally involve reducing either the *breadth* or the *depth*, or both, of the decision maker's consideration of the alternatives. For example, the decision-making heuristic called *satisficing* reduces the breadth of the search for alternatives. It takes the shortcut of setting an acceptance threshold and ending the search for alternatives as soon as one is encountered that exceeds this threshold. Imagine, for example, someone who has just moved to a distant city, is hunting for an apartment, and can't search the housing market broadly because she needs to start work immediately at her new job. If she simply picks the first apartment that is acceptable, which almost certainly is not the best one available, then she will have used the satisficing heuristic. Another decision-making heuristic, called *elimination by aspects*, reduces the depth of the decision-making task. Someone using this heuristic to find an apartment first decides what attribute is most important (say, commuting distance), sets an acceptance limit (say, no more than a 20-minute commute), and eliminates all the apartments that exceed the limit. The process is then repeated, attribute by attribute (no more than $1,000 per month; a park for jogging within five blocks) until either a choice is made or the set of possibilities is narrowed sufficiently to switch over to a thorough evaluation of the finalists. This person probably won't choose the apartment that would be best overall—she will not consider an apartment that has a 22-minute commute even if it has a low rent and a beautiful park nearby—but she has certainly diminished the time, expense, and mental effort needed to find a place to live.

Given that humans and other animals usually make decisions by drawing on a toolbox of heuristics, it is remarkable that a honeybee swarm does not use these shortcut methods of decision making and instead selects its new living quarters by taking a broad and deep look at the bee housing market. As we have seen in chapter 4, a swarm makes its decision only after its scout bees have discovered numerous alternative nest sites and have performed a multifaceted inspection of each site. In the full-size, natural swarms studied by Martin Lindauer, the mean number of candidate sites reported by scouts on a swarm was 24 (range 13 to 34), and even in the small, artificial swarms studied by Susannah Buhrman and me, the mean number was 10 (range 5 to 13). And as we have seen in chapter 3, each candidate site is evaluated with respect to at least six attributes (e.g., cavity volume, entrance height, and entrance size). Thus a honeybee swarm pursues an unusually sophisticated strategy of decision making, one that involves nearly all of the information relevant to the problem of choosing the best place to build its new nest. (Note: even a honeybee swarm is not all knowing, for even though it sends out hundreds of scouts to search for candidate nest cavities, these bees probably don't discover all the available dwelling places.) A swarm is able to be so thorough in choosing its home because its democratic organization enables it to harness the power of many individuals working together to perform collectively the two fundamental parts of the decision-making process: acquiring information about the alternatives and processing this information to make a choice. We will now look at the evidence that honeybee democracy does indeed achieve nearly optimal decision making.

Best of N?

To investigate whether a swarm's scouts usually reach agreement on the best available site, I needed to go beyond observing their dances for natural nest sites on the swarm cluster. Specifically, I needed to present them with an array of artificial nest sites that differed in quality, and I needed to do so in a location lacking natural nest sites so that the scouts would focus their attention on my artificial dwelling places. With such a setup, I could find out if a swarm's scout bees consistently choose the best nest site out of a set of alternatives—biologists refer to this as solving a "best-of-N" choice problem—or if they don't actually achieve such optimal decision making.

One can imagine various ways in which a swarm's choice of its new home could fall short of perfection. As we saw in chapter 4, a swarm's scout bees do not enter all the candidate sites into their debate at the same time, but instead do so over many hours or even a few days. If the best site happens to be presented late in the debate, its supporters might have difficulty overtaking those advocating a poorer site that was presented early on and has gained much support. Or even if the best alternative is entered into the debate at the start, it could lose out if the bees advertising it fail to tout its high quality. (How scout bees indicate a site's quality in their waggle dances will be discussed in chapter 6.) Yet another way that the best site might lose out in the scout bees' debate, even if it is reported promptly and correctly, is if this site is especially hard to find by recruits, maybe because it is far away or because it has an obscure entrance opening. Either situation will hamper the mustering of support for the best site. Given the many situations that seem like they could cause swarms to perform suboptimally in choosing a home, I wondered if swarms really are so skilled at solving the best-of-N choice problem. To find out, I needed to test their decision-making skills with controlled experiments.

Mediocrity in 15 Liters

To perform these experiments, I returned in the summer of 1997 to Appledore Island, in the Gulf of Maine, where some 20 years before I had good luck in getting swarms to be interested in artificial nest sites. In the intervening years, I had studied the bees mainly at the Cranberry Lake Biological Station, deep in the forested Adirondack Mountains of northern New York State, where flowers are sparse and bees are eager to forage from artificial food sources. Studying the bees in the north woods had been thrilling; each summer my students and I had uncovered secrets about the beautiful inner workings of a honeybee colony, particularly those that enable a colony to gather its food efficiently. Also, I had fallen in love with swimming in the lake's crystalline water, watching the northern lights glow in the midnight sky, and falling asleep to the haunting calls of loons. But by 1997 I

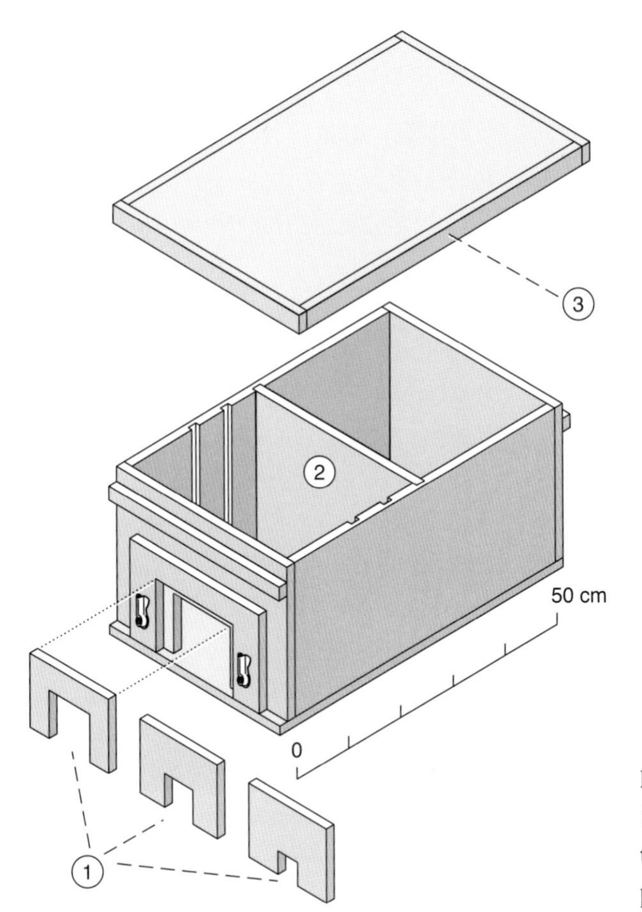

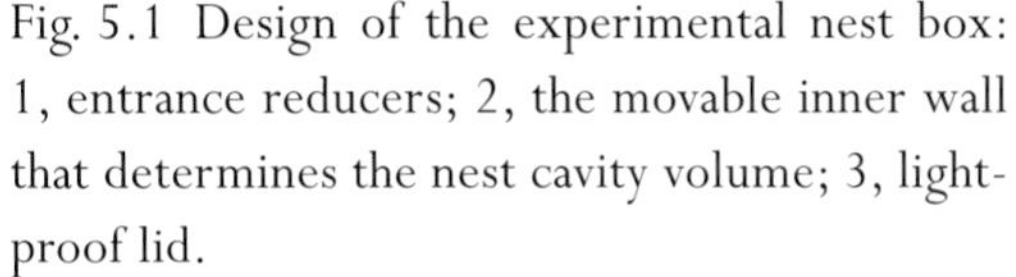
Fig. 5.1 Design of the experimental nest box: 1, entrance reducers; 2, the movable inner wall that determines the nest cavity volume; 3, light-proof lid.

was ready to return to the brilliant sunshine, ferocious gulls, lush poison ivy, and invigorating salty air of Appledore Island.

My first goal was to figure out how to make an artificial nest site that was acceptable but not ideal to the bees. If I could solve this problem, then I could test whether or not swarms achieve optimal decision making. The design of the test called for presenting swarms (one at a time) with an array of five nest boxes that would offer four acceptable homesites and one ideal homesite, and then seeing how reliably swarms would select the best of the five nest boxes. From my studies in the mid-1970s of the bees' nest-site preferences, I knew that bees prefer nesting cavities that have a large volume (40 liters) and a small entrance (15 square

Fig. 5.2 A nest box mounted in its shelter.

centimeters), so I decided to see if I could dilute the goodness of a nest box by decreasing its cavity volume or increasing its entrance area. Figure 5.1 shows the design of the nest boxes that I built. Each one had a cavity volume of 40 liters, but this could be reduced to 20, 15, or 10 liters by placing an inner wall in the appropriate location, as shown. Similarly, I could enlarge each box's entrance from 15 cm^2 to 30 or 60 cm^2 by replacing one entrance reducer with another. It was essential that the nest boxes differed only in cavity volume or entrance area, so I positioned each nest box inside an open-sided shelter (fig. 5.2). These shelters all faced the same direction so all five nest boxes had identical exposures to wind, sun, rain, and . . . gull poop.

In early August, I loaded my pickup truck in Ithaca with five nest boxes, five shelters, the swarm stand I had used for video recording the scout bees' debates, and three hives of bees for making artificial swarms. After driving to Portsmouth, New Hampshire, I loaded my equipment on the R/V *John M. Kingsbury*, the work-horse research vessel of the Shoals Marine Laboratory. It would ferry me and my 60,000 "co-workers" from the dock in Portsmouth, down the Piscataqua River, and out to the cluster of offshore islands known as the Isles of Shoals, of which the 96-acre Appledore is the largest. My 13-year-old nephew, Ethan Wolfson-Seeley, had joined me as research assistant. Soon we were off. Standing in the brilliant sunshine, drinking in the beauty of coastal New England, I felt exhilarated to be returning to one of my favorite outdoor haunts, where I had made some of my first scientific discoveries.

But I also felt slightly apprehensive to be returning to experiments with swarms, which I remembered were extremely difficult, even on Appledore Island. I had heard that Rodney Sullivan, my lobster fisherman friend, had left the island and sold his cottage. Would the new owners allow me to screen off their chimney to deter my scout bees? I also knew that over the past 20 years the Shoals Marine Laboratory had built several new dormitories and laboratories. Would these new buildings contain attractive homesites for the bees? And I wondered if I had designed my experimental nest boxes correctly, so that they could be tuned to the right settings of cavity volume and entrance area to produce a mediocre but still acceptable nest site. Would these nest boxes work? I soon stopped worrying, however, reminding myself that I've always made progress in my studies whenever I've watched the bees closely, paid close attention to unexpected results, and treated "failed" attempts to reach a goal as fingerposts indicating a better way to go forward. Certainly the remote setting of Appledore Island, 640 kilometers (400 miles) from Cornell University and 10 kilometers (6 miles) out in the Atlantic Ocean, would give me a perfect opportunity to focus my attention on the bees.

In a few days, Ethan and I had set up a swarm on the porch of one of the laboratory buildings and had placed two nest boxes in grassy sites on the north half of the island, both of them 250 meters (820 feet) from the swarm but in slightly different directions (sites A and B in fig. 5.3). To help gain the scout bees' interest, the nest boxes were set up with a large cavity volume (40 liters) and a small entrance opening (15 cm^2). Already I had introduced myself to the new owners of the Sullivan cottage (from Massachusetts and without shotgun), had explained why I wanted to put a screen over the top of their chimney, and had done so with their blessing. Now we sat patiently by the swarm, watching for bees performing waggle dances on the swarm's surface, to see what the scout bees would report. All bees announcing one of the two nest boxes would be left alone, but any bee indicating some other site would be plucked off with forceps, dropped into a small cage, and later put in a freezer. This censorship of the scout bees' communications turned out to be critical to our success. From time to time a bee would appear on the swarm dancing excitedly for a "rogue" site, and if we did not

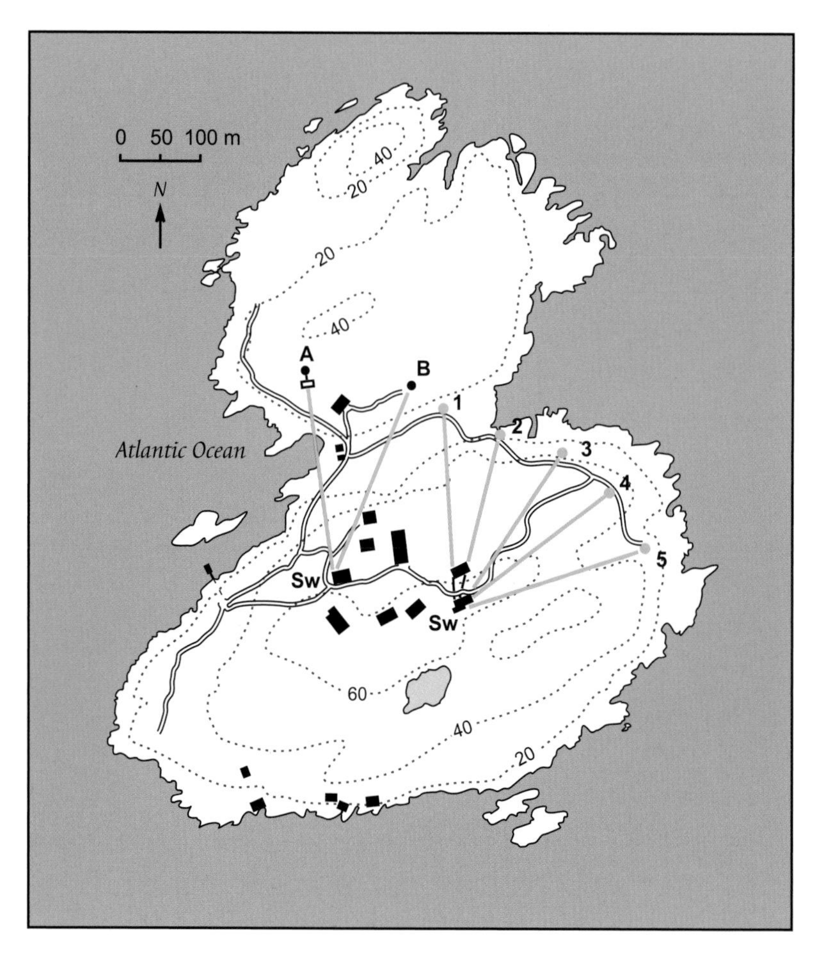

Fig. 5.3 Layouts on Appledore Island of the experiments performed in 1997 (with two nest boxes, A–B) and in 1998 (with 5 nest boxes, 1–5). Contour lines indicate feet above sea level. Sw, swarm location.

remove her quickly there would soon be an unstoppable escalation of interest in the distracting site as bees recruited to the site would come back and recruit still more bees there. Such snowballing of the scout bees' interest in an unintended site actually happened three times that summer. In two cases, we managed to find the place of interest by reading the bees' dances to determine the direction and distance to the site they were excitedly advertising, plotting its estimated location on a map of the island, and then searching there for scout bees flying in and out of some small opening. One site was a space beneath a pile of old boards and the other was a small cave in a stone wall. I rendered both sites worthless to the bees by opening them up. The third time, however, our search-and-destroy operation failed even though we poked around for hours in the right general area, which was among the three old houses on the south shore of the island. Evidently, a scout bee had discovered first-rate living quarters somewhere in the evil-looking jungle of poison ivy behind these houses, a place that we didn't dare explore. Because we could not eliminate this site, we could not extinguish the bees' raging interest in it, so all we could do was remove the swarm with its errant scout bees and start over with a new bunch of bees.

Fortunately, all our other swarms focused their house-hunting efforts on our nest boxes, and in doing so they taught us how to make one into a mediocre but acceptable dwelling place. The first lesson we learned was that I had guessed *wrong* about doing so by enlarging the entrance opening to 30 or 60 cm^2. If we gave a swarm of bees a 40-liter nest box with a 15-cm^2 entrance, they showed great interest in the box, as indicated by a rapid buildup of scout bees at the box soon after its discovery. For example, on August 10, 1997, one such nest box was found shortly before 1:00 p.m., and by 2:30 p.m. there were more than 10 bees crawling and flying about outside this nest box. There could be no doubt that the scout bees had judged this box to be highly desirable and had recruited others to it. In fact, around 1:00 p.m., back at the swarm cluster, we had observed several bees advertising the box with vigorous waggle dances. But after we enlarged the entrance opening to 60 cm^2 at 2:30 p.m., the number of bees outside the box plummeted, falling to just one or two bees by 3:00 p.m. This sudden abandonment of the box suggested that the scouts were no longer attracted to it. At 3:00 p.m. the entrance was reduced back to 15 cm^2 and the number of scout bees

Fig. 5.4 Nest box being inspected by several flying scout bees.

outside the box shot up as before, reaching more than 12 bees by 4:30 p.m. But after the entrance was enlarged again to 60 cm^2 at 4:30 p.m., the counts of the scout bees plummeted again, dropping to less than one bee by 6:00 p.m. The next day we observed the same pattern of strong buildup of bees outside the box when its entrance was 15 cm^2 and a steep crash in their number after we enlarged the entrance to just 30 cm^2. These results, confirmed by those obtained from a second swarm a few days later, taught us that scout bees judge a nest box with a 30 or 60 cm^2 entrance opening to be a low-quality, probably even unacceptable, homesite. They also showed us how easy it was to conduct an opinion poll of a swarm's scouts: simply count the bees outside each nest box (fig. 5.4).

We next tried to create a medium-quality nest site by reducing the cavity volume to something less than 40 liters. This approach worked well. Our first trial started with scout bees discovering both nest boxes late in the day on Au-

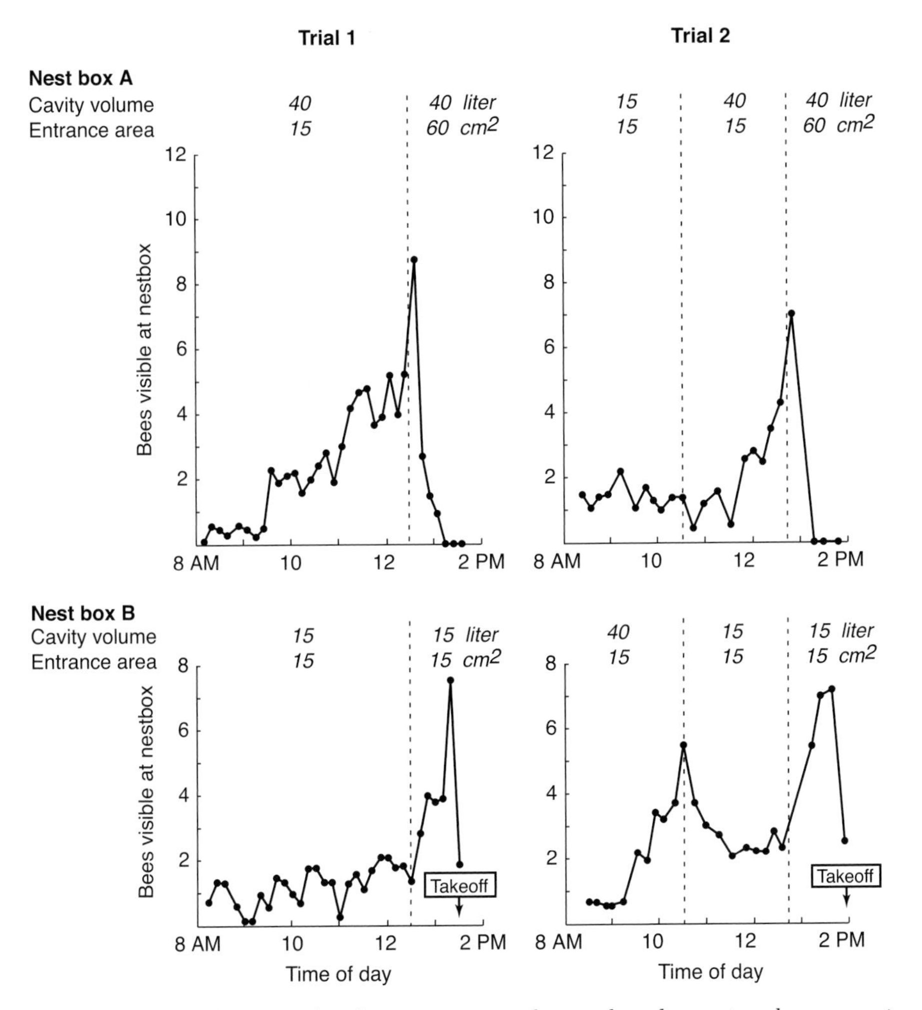

Fig. 5.5 Results of two trials of an experiment designed to determine the properties of a medium-quality nest site. In each, we presented a swarm with two nest boxes given various settings of cavity volume and entrance size. We measured the scout bees' interest in each nest box by counting the bees visible there. Dashed vertical lines show when the boxes' settings were changed.

gust 13, 1997. The next morning, we set the volume of one box at 40 liters and that of the other box at 15 liters; both boxes had the entrance opening set to 15 cm^2. As is shown in figure 5.5, the number of scout bees outside the 40-liter box rose steadily throughout the morning and reached nine bees by early afternoon.

Meanwhile, the number outside the 15-liter box stayed low at just one or two bees. It was clear that the scout bees were treating the 40-liter box as a high-quality site. But were they treating the 15-liter box as a medium-quality site, that is, one not highly desirable but certainly acceptable? To see if the bees would accept the 15-liter box, at 12:30 p.m. we enlarged the entrance opening of the 40-liter box to 60 cm^2 to render it unacceptable, and we watched to see if the bees would now accept the 15-liter box. They did! While the number of bees at the 40-liter box plunged, the number at the 15-liter box climbed to a high level and at 1:28 p.m. the swarm took off to fly to the 15-liter nest box. (The reason for the sharp drop in number of scouts at the chosen site shortly before swarm takeoff will be discussed in chapter 8.) Thus this first trial yielded evidence that we could present our bees on Appledore with a mediocre but acceptable dwelling place if we gave them one of our nest boxes with the volume set at 15 liters and the entrance set at 15 cm^2.

Additional trials made with two other swarms produced results similar to those from the first swarm. When given a choice between two nest boxes with different volume settings, a swarm's scout bees would build up much more strongly at a 40-liter box than at a 15-liter one so long as both boxes had a small (15 cm^2) entrance opening. But when the 40-liter box was severely degraded by enlarging its entrance opening to 60 cm^2, the scout bees would become numerous at the 15-liter box and eventually would accept this box for their future home.

Window on a Bee's Mind

Further evidence that we had found the right formula for creating a mediocre but acceptable homesite came from observations made at the swarm cluster rather than at the nest boxes. At the swarm cluster, we could see scout bees performing dances simultaneously for the 40-liter and 15-liter nest boxes (when both had small entrances). We could also identify which nest box each dancing bee was advertising by noting the angle of the waggle runs in her dance, for we had carefully positioned the nest boxes so that their directions differed by 30° (see fig. 5.3). (We were most grateful to the bees for saving us the trouble of giving the scouts from the two sites different labels!) Now, it is well known that when a bee performs a waggle dance to recruit hive mates to a food source, she decides how strongly she should dance based on the desirability of her flower patch. For instance, a bee advertising flowers brimming with sweet nectar might perform a strong dance that contains 100 dance circuits and lasts for 200 seconds, whereas a bee reporting on a poorer nectar source might produce a rather weak dance that contains only 10 dance circuits and lasts just 20 seconds. This correlation between flower desirability and dance strength (number of dance circuits) means that the waggle dance provides us with a "window" on a bee's mind, especially on her sense of the quality of what she is reporting to her hive mates.

Assuming that this window works for bees advertising nest sites as well as food sources, we decided to look through it to see how scout bees advertising our 40-liter and our 15-liter nest boxes judged the quality of each as a prospective home. We did so by video recording the dances performed side by side on a swarm by two groups of scouts, those reporting on our 40-liter nest box and those reporting on our 15-liter nest box. The fact that both nest boxes elicited dancing by the scout bees told us that both were of considerable interest to the bees. But even more telling was what we learned by carefully reviewing the video recordings and measuring the strength of each bee's dance. We found that bees reporting the 40-liter box performed strong dances that on average contained about 35 circuits and lasted about 85 seconds, whereas the bees reporting the 15-liter box performed weaker dances that on average contained only about 14 circuits and lasted only about 45 seconds. These findings strongly support the conclusion that the bees judged our 15-liter nest box to be an acceptable but mediocre homesite. The indication of *acceptability* is that the scouts produced dances for the 15-liter nest box (we don't expect scouts to advertise an unacceptable site), and the indication of *mediocrity* is that the scouts produced relatively weak dances for the 15-liter nest box.

Critical Experiment

On Appledore Island, during the sunny days of August 1997, I had learned from the bees how to tune my experimental nest boxes so that I could present a swarm with a choice among five possible homesites, four of them fixer-uppers and one a dream home. So at this point I was ready to present swarms with the best-of-5 choice problem, and I was extremely eager to do so, but alas I had to wait until the following summer to conduct this critical test of their decision-making skills. Fall semester classes start at Cornell in the last week of August, and each fall semester I'm part of a team that teaches a popular class in animal behavior, so I needed to get back to Ithaca to give my lectures. I also needed to set up the special glass-walled hives of bees that we use in the course to introduce students to the pleasures of watching and wondering about bee behavior.

In June of 1998, I returned to Appledore Island. With me was the smart and dedicated Cornell undergraduate student, Susannah Buhrman, who had helped me the summer before in documenting the scout bees' debates. Our goal now was to test the decision-making skills of swarms by giving them the best-of-5 choice test. Administering this test required two people working as a team, one sitting at the swarm to eliminate any scouts performing dances for sites other than the nest boxes (fig. 5.6), and one circulating among the nest boxes to count the scouts visiting them. As is shown in figure 5.3, we set up the nest boxes in a fan-shaped array on the east side of the island, so that the boxes were about the same distance (approximately 250 meters) but in different directions (at least 15° apart) from the swarm. We began each trial of the experiment by arranging the inner walls inside the five nest boxes so that one offered a 40-liter nesting cavity and the others offered a 15-liter cavity. Next, we mounted the swarm to be tested on the swarm stand. Once the swarm had formed its cluster and scout bees had begun flying from the cluster, one of us started monitoring the scouts' dances to remove reports of sites other than our five boxes, while the other person started checking the nest boxes, visiting each box every half hour and counting the bees there. We performed five trials of the experiment, each with a different swarm of bees and each with a different location for the excellent nest box. It should be noted that we did not change the location of the excellent nest box between trials by moving one excellent box around, rather we did so by leaving the five boxes in place and adjusting their volume settings. Thus in each trial a different box was given the 40-liter setting that made it the excellent option.

The full results of the five trials of this experiment are shown in figure 5.7. It shows for each trial how many scout bees were counted outside each nest box

Fig. 5.6 Susannah Buhrman censoring dancers for rogue sites on one of the swarms.

over the course of the trial. We can see that in all five trials the swarm's scouts found all, or nearly all, five nest boxes, which means that each swarm had knowledge of most of the candidate sites. We can also see that in each trial the scouts did not find the nest boxes simultaneously—though they did find them all on the same day—and that *they never found the excellent nest box first*. For example, in Trial 1, scout bees were seen at the four mediocre nest boxes in the morning but not at the excellent nest box until the afternoon. Furthermore, we can see that sometimes a substantial crowd of scout bees had formed at one or more of the mediocre boxes before even one scout had found the excellent box. In Trial 2, for example, the number of scouts outside the mediocre nest box 1 grew steadily between 11:30 a.m. and 2:00 p.m. and had reached more than five bees by the time the excellent nest box 2 was discovered, shortly before 2:00 p.m.

Given that the excellent nest box was never found first and so always started out behind in the race to gain supporters, it is impressive that in four out of the five trials (1, 2, 3, and 5) the excellent nest box eventually gained the most supporters and became the chosen site. So, the five swarms did not achieve a perfect 5-for-5 score in this choice test, but they did demonstrate impressive skill in decision making. To see why this is so, consider the probability of getting the observed outcome purely by chance. If the swarms had chosen *at random* among these five nest boxes, then the probability that they would have chosen the best box in four out of five trials is vanishingly small, just 0.0064. In other

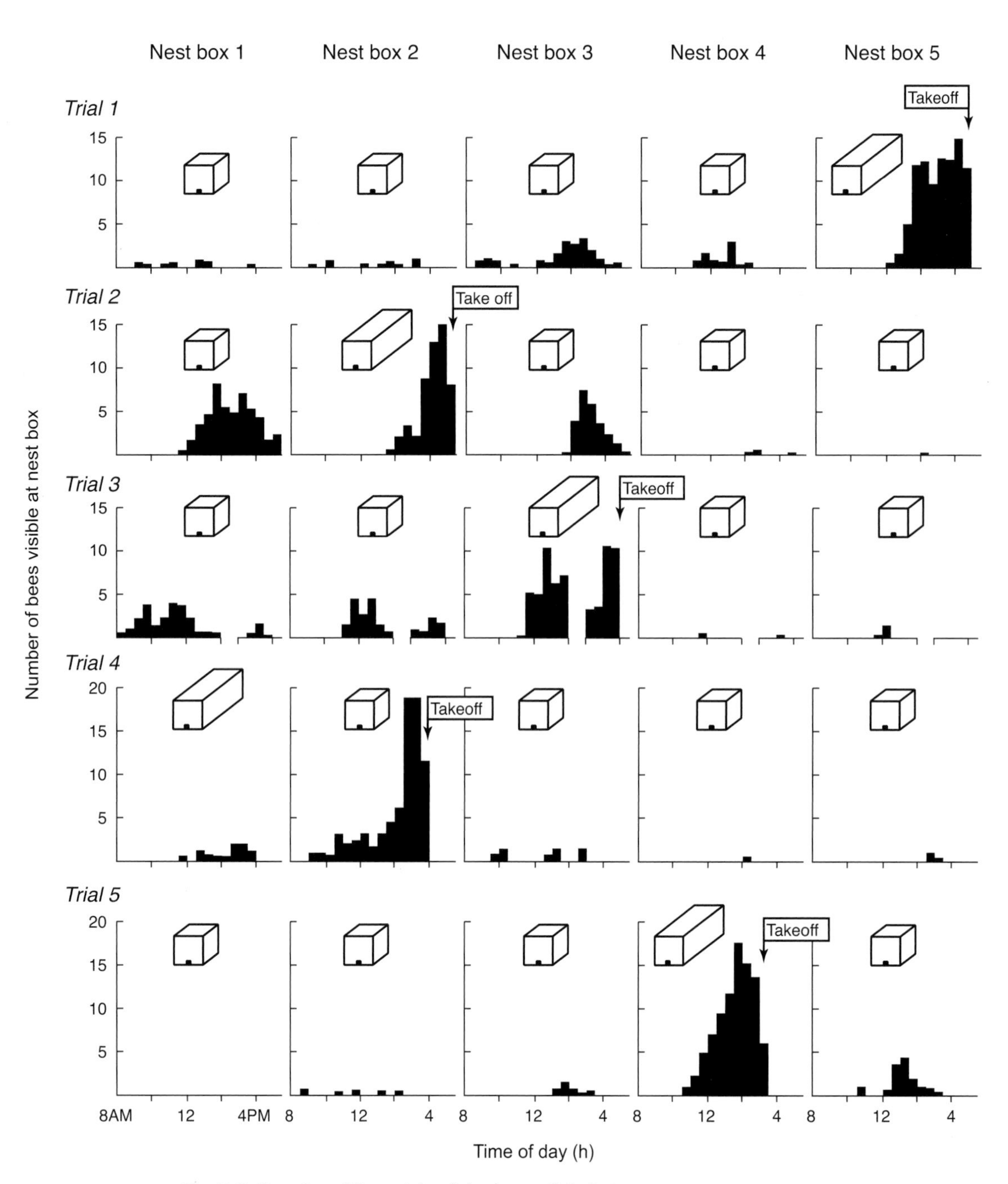

Fig. 5.7 Results of five trials of the best-of-5 choice test.

words, one would expect to get the observed outcome of four correct choices and one incorrect choice simply by chance only one time in 156 repetitions of the experiment (1/156 = 0.0064). It is clear, therefore, that compared to relying on chance, the democratic decision-making process found in a honeybee swarm greatly increases a swarm's likelihood of selecting for its future home the best of the candidate sites located by the intrepid scout bees.

I often find it useful to ponder instances where the bees behave unexpectedly, asking myself, "What is this surprise telling me?" Trial 4 of the best-of-5 choice test, in which the swarm chose a mediocre site, was a good eye-opener about how a swarm's knowledge of each prospective nest site is at first extremely fragile and easily lost. We see from figure 5.7 that in the four other trials in which the swarms chose the best site, the counts of the scout bees suddenly changed in two ways after the excellent site was found: they rose rapidly at the excellent site and they fell steadily at the mediocre sites. In Trial 4, however, neither change occurred following the discovery of the excellent site. Why not? For some reason, neither of the two scouts that discovered the excellent site ever produced a waggle dance to announce her discovery. It is puzzling that neither bee reported her find, because in Trial 2 and Trial 3 the bees that found the nest box in this location (at the north end of the array) had produced waggle dances, even though they had found only a mediocre, 15-liter nest box there. It seems clear, therefore, that there was nothing wrong about the location per se. Whatever the cause of the puzzling nondancing by scouts from the excellent nest box in Trial 4, the consequence was clear: the swarm "overlooked" the best possible dwelling place on the island. Meanwhile, a slow buildup of scout bees persisted at one of the mediocre nest boxes and eventually the swarm chose this second-rate nesting site. This anomalous outcome shows us how it is critically important to the success of a swarm's decision making that when a scout discovers a prospective homesite she reports it so that it becomes one of the options debated on her swarm. In the next chapter, we will see that the bees have a nifty rule of house-hunting behavior that normally results in every respectable housing option found by a swarm's scout bees getting included in their debate. Good decisions require good information.

Swarm Knows Best

One might question whether the results of the experiment just described really show that honeybee swarms are *good* decision makers. After all, to draw this conclusion from the best-of-N experiment, one has to assume that a 40-liter cavity with a 15-cm^2 entrance is indeed a high-quality nest site, and that a 15-liter cavity with a 15-cm^2 entrance is indeed a medium-quality nest site, so that in choosing the former over the latter a swarm improves its ability to survive and reproduce. This seems to me to be a reasonable assumption, for why would honeybees have a preference for 40-liter cavities over 15-liter ones unless natural selection has favored having this preference? Certainly, studies of other animals—including various birds, reptiles, insects, and fish—have found that the nest-site preferences of these animals enhance their reproductive success.

In 2002, I decided to test my assumption that the housing choices of honeybee swarms really are good choices, ones that help colonies survive and reproduce. Regrettably, this test required an experiment in which many colonies would die, for I needed to compare the survival probabilities of colonies living in hives embodying what the bees do and do not prefer in a home. To do this, I installed artificial swarms in hives of two different sizes in the spring, then left them alone all summer, and saw whether the two types of colonies differed in probability of surviving the following winter. (As discussed in chapter 2, most honeybee colonies living in nature starve during their first winter.) Each swarm contained approximately 10,000 bees, a typical population size for natural swarms. For the two sizes of hives, I chose ones that held either 5 or 15 of the rectangular wooden frames that hold the beeswax combs in a hive, because these are the numbers of frames needed to hold the amount of comb that bees will build inside a 15-liter or a 45-liter tree cavity. Natural swarms occupy empty tree cavities and must invest heavily in comb building, so to give my artificial swarms the same challenge I installed them in hives containing empty frames in which they had to build their combs. (I did install a sheet of beeswax "foundation" in every other frame, to induce the bees to build their combs neatly within the wooden frames.) Each year that I have conducted this experiment, I have set up five colonies of each type in early June and then followed them for the next twelve months to see which ones would survive to the following spring.

To date, I have performed three replicates of the experiment—in 2002–2003, 2003–2004, and 2004–2005—so I have followed the fates of 30 colonies. For colonies in the 15-frame hives, the probability of winter survival has been 0.73 (11 out of 15 colonies), but for the colonies in the 5-frame hives this probability has been only 0.27 (4 out of 15 colonies). The large difference in colony survival between the two treatments has only a tiny probability ($p = 0.02$) of arising simply by chance. Almost certainly, the colonies in larger hives survived better because they amassed larger stockpiles of honey to sustain them through winter. I can make this claim because I weighed each colony's hive at the start of the experiment in June and again in October, after heavy frosts ended the bees' foraging for the year, and I recorded widely different average weight gains for the two sizes of hive: 23 versus 10 kilograms (51 versus 22 pounds), most of which is honey. Also, when I examined the combs of the colonies that died in this experiment, almost always I found them empty of honey. The poor bees starved. These stark statistics on colony survival as a function of hive roominess are solid evidence that swarms really do know best about their housing needs, and in exercising their nest-site preferences they really do make good decisions. They also make clear why honeybee swarms go to so much trouble to find good homes.

6

BUILDING A CONSENSUS

We deprecate division in our Meetings and desire unanimity. It is in the unity of common fellowship, we believe, that we shall most surely learn the will of God.
—*Society of Friends,* Book of Discipline, *1934*

A dissent-free decision. This is what normally arises from the democratic decision-making process used by house-hunting honeybees and, quite frankly, I find it amazing. We have seen in the last two chapters how the debate among a swarm's scout bees starts with individuals proposing many potential nesting sites, vigorously advertising the competing proposals, and actively recruiting neutral individuals to the different camps. All this makes the surface of a swarm look at first like a riotous dance party. Yet out of this chaos, order gradually emerges. Ultimately the debate ends with *all* the dancing bees indicating support for just *one* nesting site, usually the best one. Exactly how the scout bees achieve unanimity at the end of a protracted debate is the subject of this chapter.

Consensus building is sometimes the basis of democratic decision making in human groups—such as trial juries, Quaker meetings, and groups of friends—but it is not so common. What is common is for a human group to end a debate, election, or other democratic process with its members strongly divided in their preferences. At this point the group must invoke some formal decision rule, for example, majority rule or a weighted-voting system, to translate its split vote into a single choice. This kind of group decision making has been called "adversary democracy" because it arises from a group of individuals who have conflicting interests and different preferences. In contrast, the group decision making of swarm bees is "unitary democracy" since it involves individuals who have congruent interests (choose the best homesite) and shared preferences (small entrance opening, etc.). Thus, in looking closely at the inner workings of the unitary democracy of a honeybee swarm, we will be examining a democratic process that is intriguingly different from our all-to-familiar adversary democracy. Later in the book (in chapter 10), I will discuss some practical lessons that we humans can learn from the bees for improving human group decision making, especially when the members of a group have common interests, as do the bees in a swarm.

The group solidarity with which a swarm's scout bees end their debate is critical to the success of the entire swarm. After all, a swarm contains just one queen, so when a swarm takes off to fly to its new home, it needs to do so as a single cohesive entity that travels to a single new homesite. Split decisions are wasteful and can even be fatal. As we have seen with Lindauer's Balcony swarm (see fig. 4.4), if a swarm takes off with the scouts still strongly advertising multiple homesites, the swarm won't succeed in moving to any of the sites, hence it wastes time and energy. And if the swarm loses its queen during the aerial tug-of-war between the different parties of scout bees, then it pays the ultimate price of complete failure, for it is doomed without its queen. It seems of paramount importance, therefore, that a swarm's scouts reach an agreement on just one site among the many that have been found before the swarm launches itself into flight.

A good way to begin to understand how the scout bees achieve unanimity is to reexamine the synoptic records of the scout bees' debates. Consider the debate of Swarm 3, summarized in figure 4.7. It shows two striking phenomena that must be explained to understand how a swarm's scouts build a consensus. First, there is the curious way that the support for the winning site—site G in the southwest—grew steadily and ultimately dominated the discussion. Between 1:00 and 3:00 p.m. on July 20, only 4 out of the 30 (13 percent) dancing bees advertised site G. But by 9:00 to 11:00 a.m. on July 21, 32 out of the 52 (62 percent) dancing bees advertised this site. And on the morning of July 22, shortly before swarm departure, 73 out of the 73 (100 percent) dancing bees advertised site G. Presumably, site G was the best of the 11 sites considered by this swarm, since swarms generally choose the best of the candidate homesites under consideration (chapter 5). So our first critical puzzle about the bee's system of decision making by consensus building is this: What causes the support of the scout bees for the best site to grow and grow throughout a debate?

The second striking phenomenon shown in figure 4.7 is the way that the support for all the poorer sites eventually evaporated. We can see that sometimes the loss of support happened quickly, as in the case of site A in the east. And sometimes it happened gradually, as with site B in the south. But sooner or later, all the bees that performed dances for the poorer sites lost their enthusiasm for them and ceased advertising these sites. The attrition of support for the rejected sites can also be seen at the sites themselves. Figure 5.7 shows, for example, how in the best-of-N experiment on Appledore Island the counts of scout bees at all the nest boxes except the chosen one dropped essentially to zero by the end of each trial. So our second critical puzzle about the bees' method of consensus building is this: What causes the support of the scout bees for the poorer sites to fade over the course of a debate?

Lively versus Lackluster Dances

We know that a swarm contains approximately 10,000 worker bees and that a few hundred of these bees function as nest-site scouts. We also know that a swarm's scouts locate a few dozen candidate nest sites that deserve to be advertised with waggle dances. Each candidate site is originally discovered by a single scout bee, the one who chances to find it while prospecting knotholes, crevices, and other dark places for a good nesting cavity. This means that only a few dozen scouts truly discover the sites that get debated during a swarm's decision making; most scouts learn about and become committed to a particular site by being recruited to it. Each of these recruits follows a dance advertising a site, flies out, locates the advertised site, and makes an independent evaluation. If the proposed residence satisfies her scrutiny, then she too will dance for it when she returns to the swarm.

Given these facts about scout bees, we can view a swarm's democratic choice of its future domicile as a kind of election process in which there are multiple candidates (possible nest sites), competing advertisements (waggle dances) for

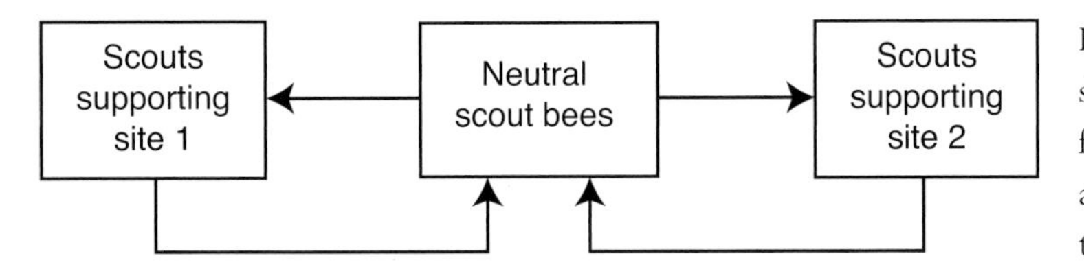

Fig. 6.1 The transitions between states that scout bees can undergo, from neutral scout to supporter for a site, and then back to being a neutral scout bee.

the different candidates, individuals who are committed to this or that candidate (scouts supporting a particular site), and a pool of individuals who are still neutral (scouts not yet committed to a site). The scouts supporting a site can produce dances that will convert neutral individuals into additional supporters for their site. Also, the scouts supporting any given site can become apathetic voters and rejoin the pool of neutral scouts. The whole decision-making process can be depicted schematically as a set of positive feedback loops of recruitment of neutral bees into supporters for the different sites, along with "leakage" of some supporters back into the pool of neutral scouts (fig. 6.1).

Looking at the scout bees' debate in this way, it is clear that in order for the supporters of the highest-quality site to be successful in ultimately dominating the debate, they must do the best job of gaining converts, presumably by showing the greatest zeal in advertising their site. Does this happen? More specifically, when an evangelizing scout bee advertises a potential nest site with a waggle dance, does she adjust the strength of her dance in relation to the absolute goodness of her site? If all the scouts do likewise, then the highest quality site should indeed receive the most compelling advertisements.

The first evidence that this actually happens comes from observations made by Martin Lindauer in the summer of 1953. He set up an artificial swarm in the broad moorlands east of Munich, and there he also set out two empty wooden hives 75 meters (about 250 feet) from the swarm. On the first day of this experiment, scouts from Lindauer's swarm quickly discovered his two hives sitting exposed in the windswept fields, and they performed rather sluggish dances advertising their two finds. Little by little, there grew a small crowd of inquisitive scouts at each hive. By the end of the first day, Lindauer had labeled 30 dancers total for his two hives. On the second day, Lindauer noticed an exceptionally lively dancer on the swarm cluster, a scout who turned out to be advertising a snug underground cavity located beneath a tree stump in the corner of a small woodlot. This site was thoroughly protected from the wind by thick bushes, had a 3-centimeter wide (1.2-inch diameter) entrance opening and a 30-liter (27-quart) cavity volume, and was wonderfully dry inside despite heavy rains in recent days. It was a perfect bee home! Lindauer normally killed all bees advertising rogue sites, but on this day he wisely made an exception; this excited bee was allowed to continue announcing her discovery. Within an hour, other boisterous dancers were also indicating the natural nest site, and after another hour, the scouts were dancing unanimously in favor of this site. It was the clear winner in this debate.

The fact that the scout bee that discovered this first-class dwelling place announced her find with an eye-catching dance, even though she had not visited either of Lindauer's test hives, suggested to him that scouts are able to judge the absolute quality of a site through reference to an innate scale of nest-site goodness. Also, the fact that this first dancer and her fellow advocates of the tree stump site danced more strongly than the bees advertising the two hive sites gave Lindauer an indication that a scout's dance provides information not only about a site's *location* but also about its *quality*. He summarized his observations by reporting: "The most lively dances indicate a nest-site of the first quality; second-rate homes are announced by lackluster dances."

Representing Site Quality in Dance Strength

Good decision making by a honeybee swarm depends critically on the scouts adjusting dance strength in relation to site quality, so that scouts advocating higher quality properties are better at attracting additional supporters. Nevertheless, it was not until the summer of 2007 that I looked closely at how nest-site scouts provide information about site quality in their waggle dances. I had recognized for years that Lindauer had made only preliminary observations on this important subject, so I had long known that more convincing evidence was needed, but I had procrastinated.

I left this gaping hole in the analysis open for so long because I had little doubt that what Lindauer had claimed was correct: better sites elicit stronger dances. It

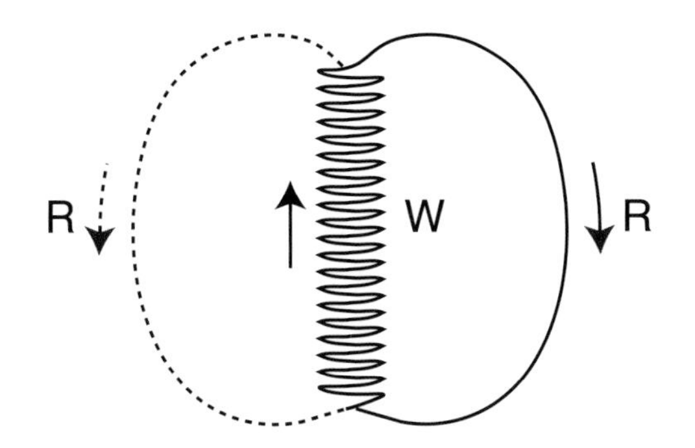

Fig. 6.2 Movement pattern of a bee performing the waggle dance. Each dance consists of a series of dance circuits. Each dance circuit contains a waggle run (W) and a return run (R, alternating right and left). The duration of the waggle run depends on the distance to the target (food source or nest site). The duration of the return run depends on the desirability of the target. As target desirability increases, return run duration decreases, making the dance appear livelier.

was certainly consistent with what I had observed here and there. For example, I had often noticed how some scout bees perform longer and livelier dances than others. Also, from the best-of-5 choice test conducted on Appledore Island where I had seen scouts performing dances, side by side, for either a 40-liter or a 15-liter nest box (see Window on a Bee's Mind, in chapter 5), I had seen that the bees reporting on the better homesite performed stronger dances. Furthermore, in previous studies by myself and others on how a honeybee colony wisely deploys its foragers among nectar sources—a group decision-making process that depends on a colony's foragers making graded advertisements of the various nectar sources—we had found that the richer the nectar source that a bee exploits, the greater the number of dance circuits she produces when she returns to the hive and advertises the source. In short, the richer the nectar source, the stronger the waggle dance. We had also figured out how a dancing bee adjusts the number of dance circuits that she produces in relation to nectar-source richness. She does so by adjusting two aspects of her dancing: the *rate* of dance circuit production (R, in dance circuits per second) and the *duration* of dance circuit production (D, in seconds) (see fig. 6.2). The total number of dance circuits produced (C, in dance circuits) in a dancing bee's advertisement is the product of the rate and duration of her dancing ($C = R \times D$). So, richer nectar sources elicit livelier (higher R) and longer-lasting (greater D) dances than do poorer nectar sources. These findings about nectar-source foragers matched perfectly with Lindauer's report that nest-site scouts announced an inferior nesting place with a "faint-hearted dance" while those from a superior nesting place "solicited with a lively and long-lasting dance."

By 2007, however, I had reached the point in my analysis of the bees' house-hunting process where I knew that I really needed to get solid, quantitative information about how the scouts code nest-site quality in their dances. To accomplish this, we would need to work under the controlled conditions provided by Appledore Island. I say *we* because I teamed up with two collaborators on this project, Marielle Newsome, an undergraduate student at Cornell, and Kirk Visscher, a behavioral biologist from the University of California at Riverside. Marielle had done beekeeping with her father and was headed to graduate school at the University of Michigan to study insect behavior, so she was keen to help. Kirk is a longtime collaborator in various bee studies, going back to when we were both students at Harvard, and he has always been the best possible partner: intelligent, skilled, good-natured, and highly enthusiastic.

Our plan called for positioning one artificial swarm in the center of Appledore Island and two nest boxes 250 meters (820 feet) from the swarm but only about 40 meters (130 feet) apart so the swarm's scouts would be likely to find the two boxes more or less simultaneously (fig. 6.3). One box offered a high-quality (40-liter) nesting cavity while the other presented a medium-quality (15-liter) one. For each of the first five to seven scout bees that appeared at each box, we would record with a data logger when she was at "her" nest box and we would

Fig. 6.3 Marielle Newsome recording visits by individual scout bees to a nest box housed in the orange shelter. In the background, 40 meters away, Kirk Visscher is doing the same at the second nest box.

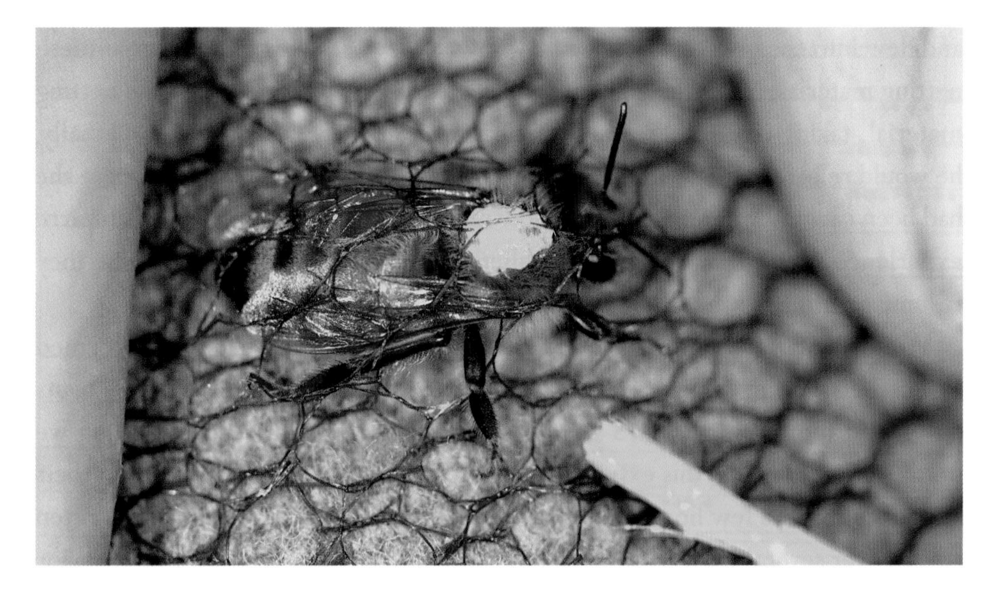

Fig. 6.4 A scout bee that has been labeled with a paint dot on the thorax, applied through the coarse-mesh netting of an insect net.

record with a video camera when she was at the swarm and how strongly she danced to advertise her site. Analysis of the video recordings in the evening would reveal exactly when each scout danced and how many dance circuits she produced. What made the execution of this experiment seem daunting at first was the fact that to examine the behavior patterns of individual scout bees, we would need each scout to be individually identifiable as soon as she was sighted at one or the other nest box. I expected this need would require us to laboriously prepare swarms in which each bee was labeled for individual identification (see fig. 4.5). We certainly had no way to know in advance which of the thousands of bees in a swarm would first appear at our nest boxes, so we couldn't label in advance just the few first pioneering scout bees.

Fortunately, Kirk had an ingenious solution to our scout bee ID problem. In a previous study, he had found that he could apply identifying paint marks to a scout bee during her visit to a nest box without distressing her. To do so, he placed a small insect net over the box's entrance after he saw a scout bee go inside to inspect the box's interior. Then, when the scout came out a minute or so later

tial for decision-making errors arising from individual-level noise in reporting on sites is especially great when each site is discovered, for if the scout that discovers a site fails to report on it with a waggle dance, the site won't be entered into the scout bees' debate. Indeed, it will be lost from the swarm's attention unless another scout happens to find and report the same site, which is most unlikely. A solution to this problem would be to have each scout bee that discovers a site likely to report on the site and thereby enter it into the debate. Marvelously, the bees appear to do exactly this. In our experiment, Marielle, Kirk, and I found that the two scout bees that first visited the two nest boxes in each trial almost always (with probability of 0.86) performed waggle dances upon return to the swarm, whereas the scouts that visited the same nest boxes subsequently, probably having been recruited to the boxes, were somewhat less likely to perform waggle dances (with probability of 0.55). We do not know what gave the initial scouts their especially strong motivation to dance. Perhaps it was each initial scout's experience of finding the site by herself—not having followed another scout's dance to find it—or of inspecting the site by herself. This "discoverer-should-dance" rule is not foolproof, however. As we have already seen in the best-of-5 choice test, in which swarms were presented with a five-alternative choice (one 40-liter nest box and four 15-liter nest boxes), one swarm failed to choose the best 40-liter option because two scout bees that discovered it independently both failed to report it with dances (see fig. 5.7). Consequently, the swarm "overlooked" the excellent alternative and ended up occupying one of the mediocre ones.

There is one more important feature of scout bee behavior that caught our attention in July 2007: each of the marked scouts visited *just one* of our two nest boxes even though the two boxes were only 40 meters (130 feet) apart, a distance that a flying bee needs only 10 seconds to traverse. Such site fidelity by the scouts is noteworthy because it provides further support for Lindauer's suspicion that when a scout bee evaluates a prospective homesite, she makes an estimate of its absolute quality based on an innate (genetically specified) scale of nest-site goodness. In other words, she does not make an estimate of a site's relative quality by comparing it to other sites that she has visited. Because our swarms were prepared from colonies that had not recently swarmed, we could be sure that our bees had no prior experience as scouts before coming to Appledore Island.

And because we did not see any of them visit more than one nest box on the island, we could be confident that they did not compare one site with the other. Nevertheless, those that visited the high-quality site danced more strongly than those that visited the medium-quality site. Evidently, a worker bee possesses both an innate knowledge of what constitutes an ideal homesite and an innate ability to determine the absolute quality of the site that she has inspected. This is not a far-fetched claim; various studies of worker honeybees have shown that when a flower-naive bee searches for flowers, she spontaneously prefers objects with complex shapes, certain colors (e.g., violet rather than green), and certain odors (floral rather than nonfloral). This innate knowledge of floral cues naturally steers the novice forager's attention toward flowers.

Finally, I should emphasize that almost certainly a scout bee does not consciously think through her evaluation of a site. Instead, she probably does so unconsciously with her nervous system integrating various sensory inputs relating to cavity size, entrance height, and the like, and generating within her a sense of the site's overall goodness. It may be that finding a desirable tree cavity feels to a homeless scout bee as inherently pleasurable as feasting on a delicious meal does to a hungry human being.

The Strong Grow Stronger

One key to understanding why the scout bees' support for the best site grows and grows throughout a debate is that the supporters of the best site advertise it the most strongly. To be precise, the scout bees from the best site produce the greatest number of dance circuits per bee, on average, as we have just seen (fig. 6.6). And this is true in nature, not just in experiments. Consider again the scout bee debate depicted in figure 4.7, in which site G to the southwest prevailed, presumably because it was the best available site. Throughout the debate, the bees advertising site G produced the greatest number of dance circuits per bee. For example, between 3:00 and 5:00 p.m. on July 20, when there was fierce competition among sites A, B, D, and G, the average numbers of dance circuits produced per scout bee for these four sites were 59, 29, 42, and 74. Likewise, the next morning, between 9:00 and 11:00 a.m., when the contest had narrowed to sites

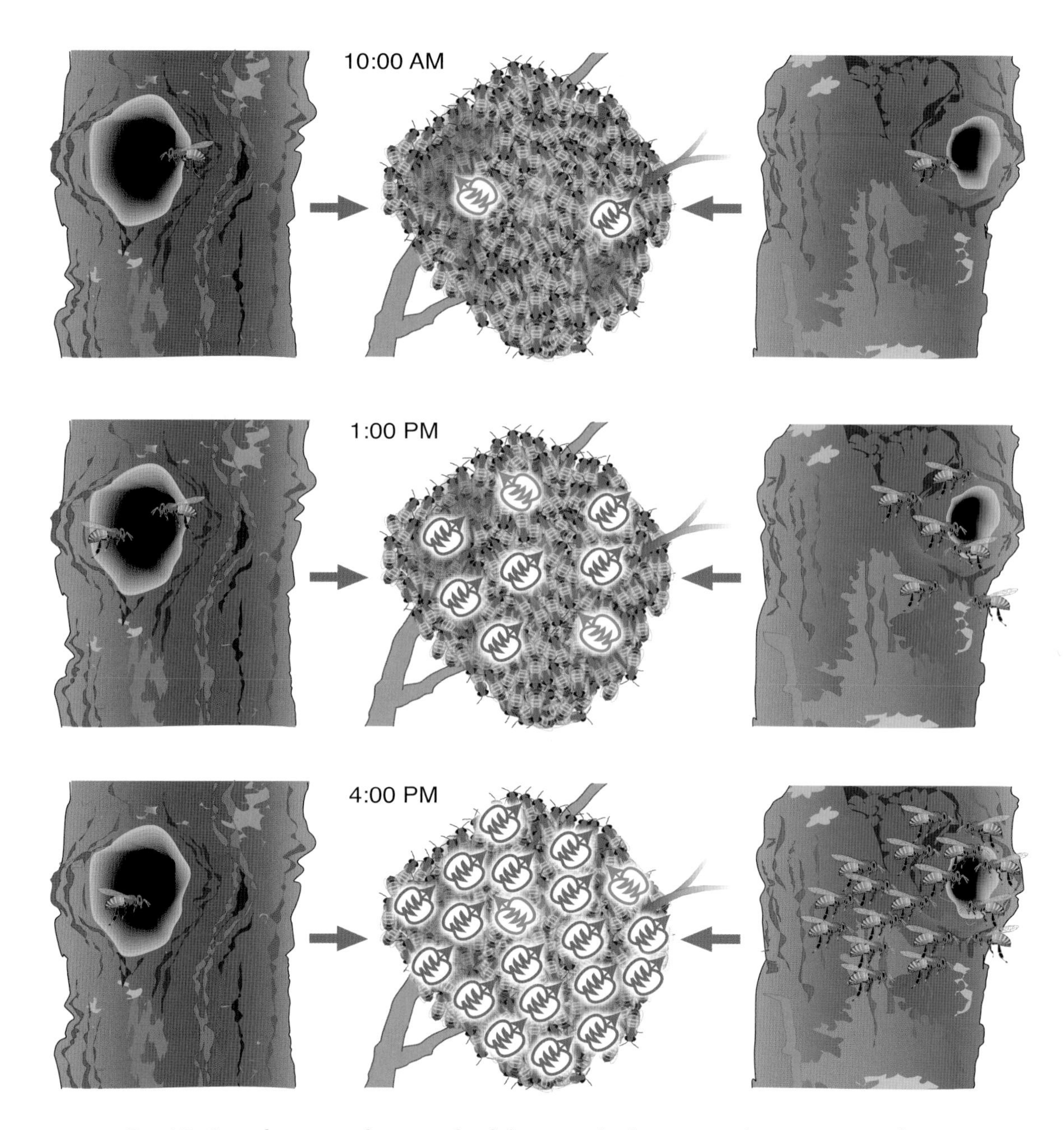

Fig. 6.7 Scout bees tune the strength of their waggle dancing in relation to site quality, which builds a consensus of dancing bees for the best site. Here, two scouts simultaneously discover two potential nest sites, one with a large entrance opening (left) and one with a more desirable small opening (right). Each scout then returns to the swarm and performs a waggle dance for her site, but the scout from the right tree performs three times as many

B and G, the average numbers of dance circuits produced per scout bee for these two sites were 16 and 42. (Note: the bees danced only about half as strongly this morning relative to the previous afternoon because the weather had deteriorated overnight. Indeed, a rainstorm started at the end of the morning. The bees always slow their house-hunting process in cool or stormy weather.)

Because the best site stimulates its supporters to dance the most strongly, its supporters have the highest per capita success in converting neutral scouts into additional supporters. And because these additional supporters will likewise have the greatest per capita success in attracting still more supporters, the differences in number of supporters among sites that differ in quality will grow exponentially. In principle, one group of supporters will eventually overwhelm all the others, which is precisely the pattern that we have seen in the swarm bees' debates (see figs. 4.6 and 4.7).

Figure 6.7 illustrates how this works for the basic situation of two competing sites that differ in quality. The high-quality site on the right, which is more desirable by virtue of its smaller entrance opening, stimulates its supporters to advertise it with 90 dance circuits on average (as did our 40-liter nest box; see fig. 6.6). The medium-quality site on the left, which has a larger entrance opening, elicits 30 dance circuits on average from its supporters (as did our 15-liter nest box). The two sites are discovered simultaneously, each by just one scout, at 10:00 a.m. During the first three hours, the two scouts produce 90 dance circuits and 30 dance circuits, so the relative force of persuasion (total amount of advertising) for the two sites is 3:1. If we assume that 8 neutral scouts are recruited to the two sites, and in proportion to the level of advertising for each site, then by 1:00 p.m. there will be six scouts supporting the high-quality site and two supporting the medium-quality one. (By 1:00 p.m. the two original scouts

---

waggle dance circuits (blue symbol) as the scout from the left tree (red symbol). The result is that three hours later, the number of bees committed to the right tree has increased sixfold, whereas support for the left tree has increased only twofold, and the majority of dancing bees favor the right tree. After three more hours, the number of scouts at the right tree has ballooned, and the numerous dances in support of this site have nearly excluded the left-tree site from the debate.

will have ceased advertising and visiting the sites.) Now what happens over the next three-hour period? The six supporters of the high-quality site will produce a total of 540 dance circuits (six bees x 90 dance circuits per bee) while the two supporters of the medium-quality site will produce a total of 60 dance circuits (two bees x 30 dance circuits per bee). Thus the relative force of persuasion for the two sites becomes 9:1 during this second three-hour period. If 20 neutral scouts are recruited to these sites (more recruits now than before because there is more advertising), and if they are recruited to the two sites in proportion to the amount of advertising for each, then by 4:00 p.m. there will be 18 scouts supporting the high-quality site but still only two supporting the medium-quality site. So we can see that even though this debate started out with a 1:1 ratio of supporters for the two sites, after three hours the ratio became 3:1, and after three more hours it reached 9:1. We can also see that if the debate continues, it won't be long before the high-quality site achieves complete domination of the debate, just as in nature.

A curious feature of the bee's consensus-building process is that the domination of the debate by one site's supporters can be driven entirely by differences in the per capita strength of advertising of the various sites. One might suppose that building a consensus among the dancing bees would also require the neutral scout bees that are getting converted into supporters to pay attention to the different types of advertisements and ignore the weaker ones representing poorer sites. But in fact, the neutral scouts don't need to follow dances selectively. In the example just given, the neutral bees become supporters for the two sites strictly in proportion to the amount of dancing for the two sites. It is as if a neutral scout simply strolls across the surface of the swarm, follows the first dance that she encounters, gets recruited to the site advertised by this dance, and then becomes a supporter for this site. Although we don't know if this is exactly how a dance-following scout bee behaves, we do have evidence that they do not selectively follow dances for certain sites but instead follow dances at random.

The evidence comes from an experiment conducted by Kirk Visscher and one of our mutual friends, Scott Camazine, a gifted physician, nature photographer, and fellow honeybee fanatic. In December 1995, in the desert east of Indio, California, where large trees are rare and so natural homes for honeybees are scarce,

Kirk and Scott set up artificial swarms (one at a time) and two nest boxes. These boxes attracted the interest of the scout bees from their swarms. Kirk and Scott then labeled for individual identification each scout that performed a dance for one of the nest boxes, and they video recorded all instances of dancing and dance following throughout each swarm's decision-making process. Then they reviewed the recordings to see which of their labeled dancers eventually became dance followers. For those that did, they determined whether each bee selectively followed dances for the nest box that was not the one she had previously visited and advertised, perhaps so she could do some "comparison shopping." Remarkably, they found that the dancers that became dance followers followed dances for the two nest boxes simply in proportion to the amount of dancing for the two boxes. Thus these bees gave no sign of doing anything more sophisticated than following dances chosen at random.

We see, therefore, that the debating scout bees appear to use a simple method to build an agreement: the better the potential homesite, the stronger the dancing of the scout bees supporting it and the greater their effectiveness in recruiting additional supporters for their place. The new supporters of each spot visit and evaluate it for themselves—thereby checking the "claims" of the previous advocates of the site and avoiding untested information being spread like a rumor—and then they likewise announce it with dances, weak or strong according to their evaluations of the place. Bit by bit, because the positive feedback (the recruitment of recruiters) is strongest for the best site, the supporters for this site increasingly dominate the discussion. Complete agreement requires, however, not only that the support for the best site steadily grows, but also that the support for the poorer sites gradually fades. We will turn now to seeing how the support for the losing sites melts away.

The Expiration of Dissent

For an agreement to emerge within a group that is debating multiple options, all of the group's members who start out supporting the losing options must eventually withdraw their support for these options and either switch their support to the winning option or quit the debate altogether. In short, the dissent must expire. We have seen that this happens in the dance debates among scout bees on honeybee swarms (see figs. 4.6 and 4.7), such that every bee that starts out dancing for a rejected site eventually ceases doing so, but we've not yet seen exactly how this occurs. Back in the early 1950s, Lindauer wrestled with this important puzzle about the bee's consensus-building process but he never quite solved it. He seemed to favor the idea that a scout bee ends her support—her dancing—for one site only when she learns about a superior site and shifts her dancing to it. He expressed this view as follows:

> Scout bees that could only find lesser nest sites easily change their votes in favor of a different nest site. Even if they dance for "their" nest site at first, they decrease their dancing bit by bit, become noticeably more interested in the lively dances of the other scout bees and finally take off to seek out the other nest site. On their inspection visits they can now draw a comparison between their own and the new nest site and, if the latter is really more suitable, from now on they also dance for it on the swarm cluster. In this way all the interest of the scout bees is concentrated bit by bit on the best of all the nest sites.

There are two critical elements of this hypothesis for how scouts cease dancing for losing sites: a bee *compares* her old site with a new site (to which she was recruited by lively dances of other bees), and if she finds the new site superior she *converts* to dancing for the new and better site. Thus we can call this the compare-and-convert hypothesis for the expiration of dissent. It is certainly a plausible hypothesis. It is, after all, how we humans usually resolve disagreements in a debate; the group's members propose various courses of action, individuals hear and compare the various proposals, and eventually the individuals who initially favored a losing proposal change their minds and convert to supporting the winning proposal. I suspect that Lindauer reasoned by analogy to consensus building by humans as he worked to understand how the scout bees reach an agreement, for he described the bees as not remaining "stubborn about their first decision" and letting "their minds be changed."

Even though Lindauer stressed the compare-and-convert hypothesis to explain the expiration of dissent among nest-site scouts, he also reported some observations that weren't entirely consistent with this hypothesis. For example, he wrote, "It is still not understood why those scout bees that had found an inferior nest site gave up dancing for the site over time, even when nothing changed about their nest site and they had not yet inspected any new housing possibilities." Clearly, he had seen instances in which a scout bee quit dancing for one site even before she knew about another site, hence before she could compare her old site against a new one. Indeed, in his magnum opus of 1955 Lindauer included a beautifully detailed record of one scout who quit dancing for one site and then sat quietly on the swarm for nearly two hours before she began following dances that directed her to a second site (fig. 6.8). This shows clearly that sometimes a scout bee will quit dancing for a site without first comparing it to another site.

The two critical elements of this alternative hypothesis for how scouts cease dancing for losing sites are these: a bee *does not compare* her old site with a new site, and she *does not convert* to dancing for a new and better site. Instead, she simply loses her motivation to dance for one site and then becomes quiescent, not even visiting her site. Thus we can call this the retire-and-rest hypothesis for the expiration of dissent.

Whenever you have two competing and mutually exclusive hypotheses to explain a single mystery, you can determine which one is false by identifying some phenomenon about which the two hypotheses make clearly different predictions. You then go out, observe the critical phenomenon, and see which hypothesis doesn't correctly predict what you have observed. You know immediately that this hypothesis is false. This "strong inference" procedure may sound esoteric, but it is something we all do all the time. For example, if the light doesn't go on in a room when you turn on the light switch, you wonder if the cause is (hypothesis 1) the bulb burned out or (hypothesis 2) the power went out. If the former, then you predict that the lights will work in another room, but if the latter, then you predict that they won't. So you check the lights in another room and when you find that they work, you know immediately that the power-went-out hypothesis is bogus.

To distinguish between the compare-and-convert hypothesis and the retire-and-rest hypothesis for how the dissent among the dancing scout bees expires, I made use of the fact that these two hypotheses make distinctly different predictions about when a scout will cease dancing for a losing site relative to when she follows a dance for another site. A critical prediction of the compare-and-convert

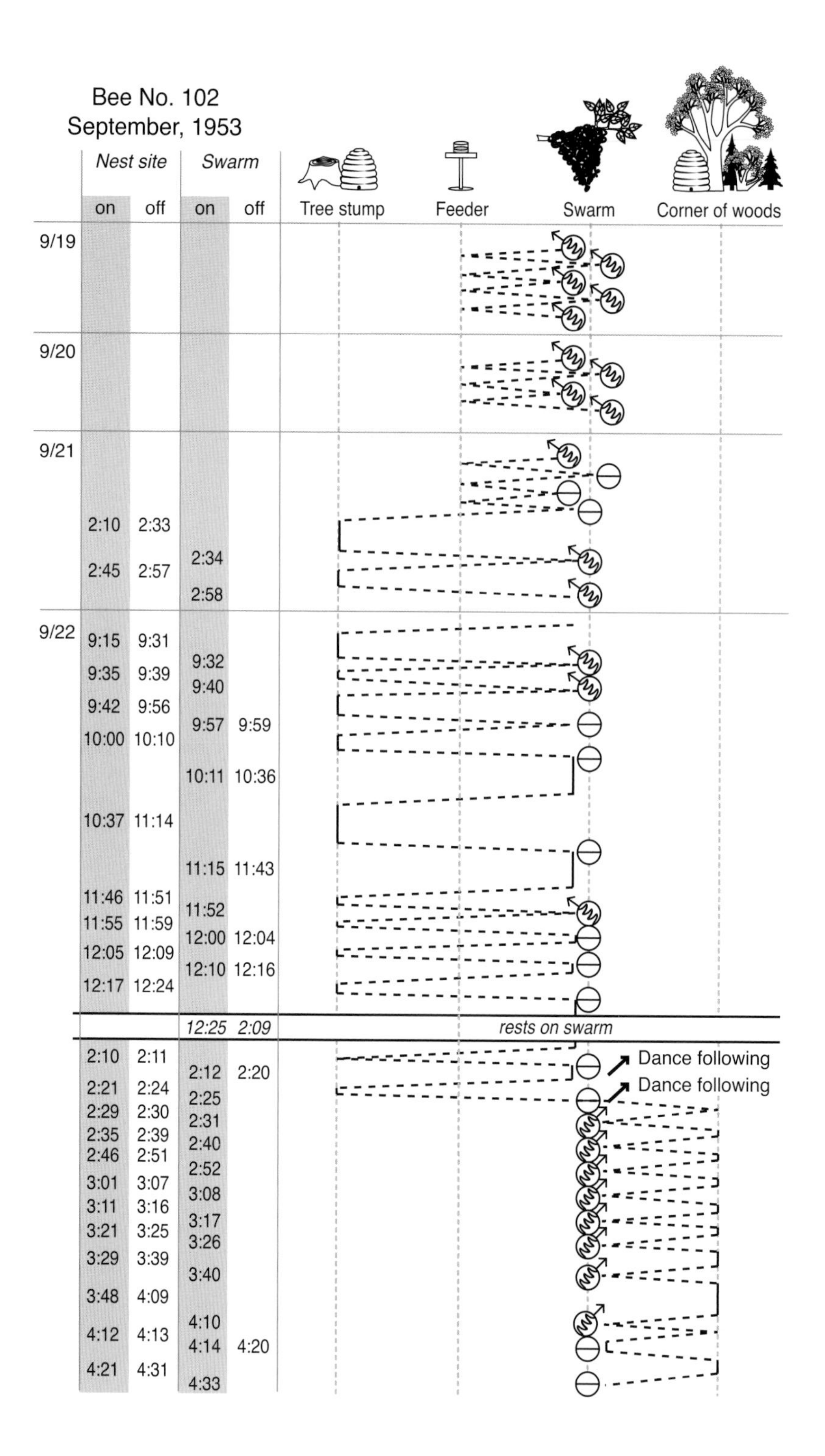

Bee No. 102
September, 1953
Nest site
Swarm
on
off
on
off
Tree stump
Feeder
Swarm
Corner of woods
9/19
9/20
9/21
2:10 2:33
2:34
2:45 2:57
2:58
9/22
9:15 9:31
9:32
9:35 9:39
9:40
9:42 9:56
9:57 9:59
10:00 10:10
10:11 10:36
10:37 11:14
11:15 11:43
11:46 11:51
11:52
11:55 11:59
12:00 12:04
12:05 12:09
12:10 12:16
12:17 12:24
12:25 2:09
rests on swarm
2:10 2:11
2:12 2:20
Dance following
Dance following
2:21 2:24
2:25
2:29 2:30
2:31
2:35 2:39
2:40
2:46 2:51
2:52
3:01 3:07
3:08
3:11 3:16
3:17
3:21 3:25
3:26
3:29 3:39
3:40
3:48 4:09
4:10
4:12 4:13
4:14 4:20
4:21 4:31
4:33

hypothesis is that a scout will cease dancing for a losing site *only after* she has followed a dance for another site (and then located this site and compared it to her current site). In contrast, a critical prediction of the retire-and-rest hypothesis is that a scout will cease dancing for a losing site *even before* she has followed a dance for another site. Testing these two predictions was simply a matter of setting up swarms one at a time, labeling with bright paint marks the first few bees that performed dances on each swarm, and then observing these labeled bees steadily whenever they were at the swarm to see when they danced and when they stopped dancing, and when (if ever) they followed the dances of other bees. I focused my attention on the first few dancers to appear on each swarm because I knew from eavesdropping on the scout bees' debates that the early dancers tend to advertise losing sites.

Since I needed to be able to observe all instances of my focal bees producing or following dances, I limited myself to labeling only a few (four to eight) scout bees on each swarm. This, in turn, meant that I needed to repeat the entire observation protocol on several swarms to get data on a sufficient number of bees. The work would be slow going, but this was fine by me. I knew that it would be both pleasurable and valuable to watch steadily my small company of brightly colored scout bees on a swarm—noting for each individual all her comings and goings, and all her dance producings and dance followings—until the swarm finished choosing its new home. Times spent outdoors closely observing bees always include the thrill of discovery.

I watched 37 scout bees in six swarms, which required a total of 66 hours of steady observation. As expected, most (31, or 84 percent) of the scouts first

---

Fig. 6.8 Records from the life of swarm bee Number 102, who served initially as a forager, then became a scout bee that first advertised a nest site (empty hive) beside a tree stump and later switched to advertising a different nest site (empty hive) at the edge of a woods. Dashed lines show flights to or from the swarm. Solid lines denote times spent at the swarm or one of the nest sites. A circle with a wavy line indicates a dance, and the arrow indicates to which feeder or nest site the dance referred. The swarm was not well fed when it was set out, so some bees (like Number 102) foraged from a feeder at first, then they became sluggish foragers as the swarm became well fed, and finally they began scouting for nest sites.

advertised a site that was eventually rejected and only a few (six, or 16 percent) danced initially for the site that was ultimately chosen by their swarm to be its future home. Of the 31 bees that started out supporting a losing site, 27 ceased advertising their sites before the end of the swarm's decision making and the other four almost did so, for their dancing had become feeble by the time the swarm finished its decision making. The key question, then, is how did the 27 bees that quit supporting a losing site do so? Did they stop dancing only after or even before they had followed dances for other sites? Figure 6.9 shows how three of these bees behaved on one swarm that chose a site in the south for its new home. We see that the first bee, Red, stopped dancing for a losing site in the west on her second trip back to the swarm, and that she did so without first following a dance for another site. Likewise, the second bee, Pink, stopped dancing for a losing site in the southwest on her third trip back to the swarm, and she too did so without first following a dance for another site. It was not until her fourth trip back to the swarm that she followed five circuits of a dance promoting a site to the west, and so possibly learned about an alternative site. Finally, the third bee, Orange, stopped dancing for a losing site in the east on her fifth trip back to the swarm, and just like Red and Pink she did so without first following a dance for another site. So all three of these bees ceased their dancing *before* they followed a dance for another site. Their behavior was typical. Of the 27 bees of interest, 26 (96 percent) stopped dancing for their losing sites *before* they followed dances for other sites and only one (4 percent) stopped her dancing for a losing site *after* she had followed a dance for another site. The finding that only one out of the 27 bees stopped her dancing for a losing site after she had followed a dance for another site indicates that the compare-and-convert hypothesis is incorrect, at least for the vast majority of scout bees. These results also increase our confidence that the retire-and-rest hypothesis is correct.

So what caused the dancers for the losing sites to retire from advertising these sites? Clearly, most were not stimulated to do so by encountering a bee dancing extremely enthusiastically for another site, for most ceased producing dances before they followed any dances. One strong possibility is that the bees were driven to retire from advertising the losing sites by an internal, neurophysiological process that causes every scout to gradually and automatically lose her motivation

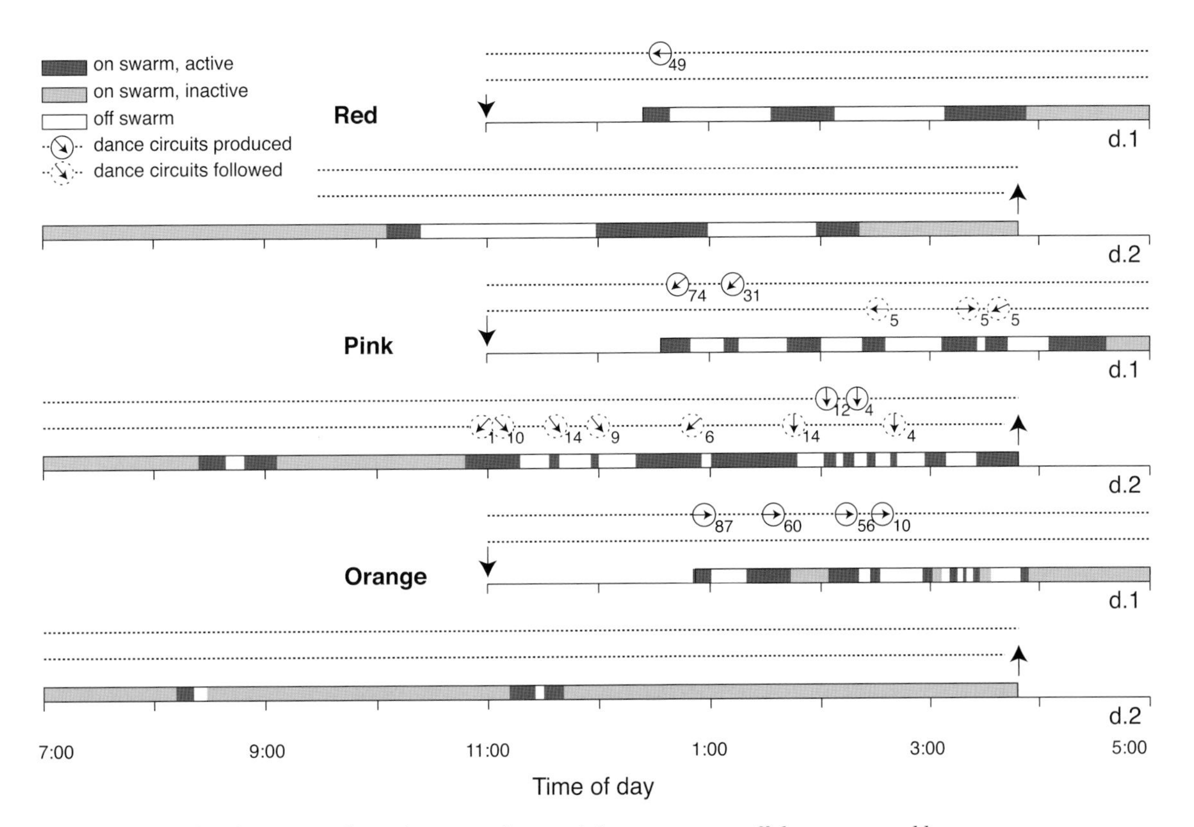

Fig. 6.9 Plots for three scout bees showing when each bee was on or off the swarm and how much dancing she produced or followed each time she was on the swarm. Each bee's history is shown for the two days over which the swarm chose its future home. The large arrows at the start and end of each bee's record denote when the swarm settled and lifted off. Each circle enclosing a small arrow denotes a dance that a bee produced or followed, and the arrow's direction indicates the compass direction of the site (an arrow pointing straight up means north, etc.). The number beside each circle enclosing an arrow shows the number of dance circuits that the bee produced or followed.

to dance for a site, even one that is high in quality. Such a process would foster consensus building among the scouts, for automatic fading of each bee's dancing would prevent the decision making from coming to a standstill with groups of unyielding dancers deadlocked over two or more sites. It might also help the dancers reach unanimity more quickly than they would otherwise, for endowing each bee with an automatic tendency to lose interest in any given site would make each bee a highly flexible participant in the decision-making process.

One piece of evidence that strongly supports the idea that scout bees have an internally driven tendency to stop dancing for any given site is something I noticed about the 37 scout bees that I watched to test the compare-and-convert and the retire-and-rest hypotheses: each bee reduced the strength of her dancing over consecutive trips back to the swarm. For example, in figure 6.9, we see that for the bee Red the decline in dance strength (number of dance circuits per trip to the swarm) was abrupt: 49 then 0. For the bees Pink and Orange, however, the declines in dance strength were more gradual: 74, 31, then 0; and 87, 60, 56, 10, then 0. (Note: one can also see this consistent drop in dance strength in the dance records of the individual scouts shown in figure 6.5.) When I tabulated all instances, for all 37 scout bees in which a bee made a series of returns to the swarm with dancing for a particular site followed by a return without dancing, I found that the bees had produced 51 such series. They varied in length from one trip back with dancing to six consecutive trips back with dancing. Then I grouped the 51 series into six sets according to series length, and for each set I calculated the mean number of dance circuits in trip 1, in trip 2, and so forth. Finally, I compared the results for the six sets by aligning them with respect to the trip back when the scout bee did not dance, as shown in figure 6.10. This revealed that, regardless of series length, there was a regular pattern of the scouts producing fewer and fewer dance circuits across a series of trips back to the swarm, and that the rate of decline in the number of dance circuits per trip did not differ markedly between bees producing long and short series. On average, there is a remarkably regular decline in the number of dance circuits produced per trip back to the swarm, and the rate of this decline is approximately 15 fewer dance circuits per trip.

It is important to note that the same pattern of steady decay in dance strength is seen with all scout bees, both those advertising a chosen site (high in quality) and those advertising a rejected site (lower in quality). The only difference is that a bee that advertises a high-quality site tends to start her reporting by performing a large number of dance circuits, whereas one that advertises a low-quality

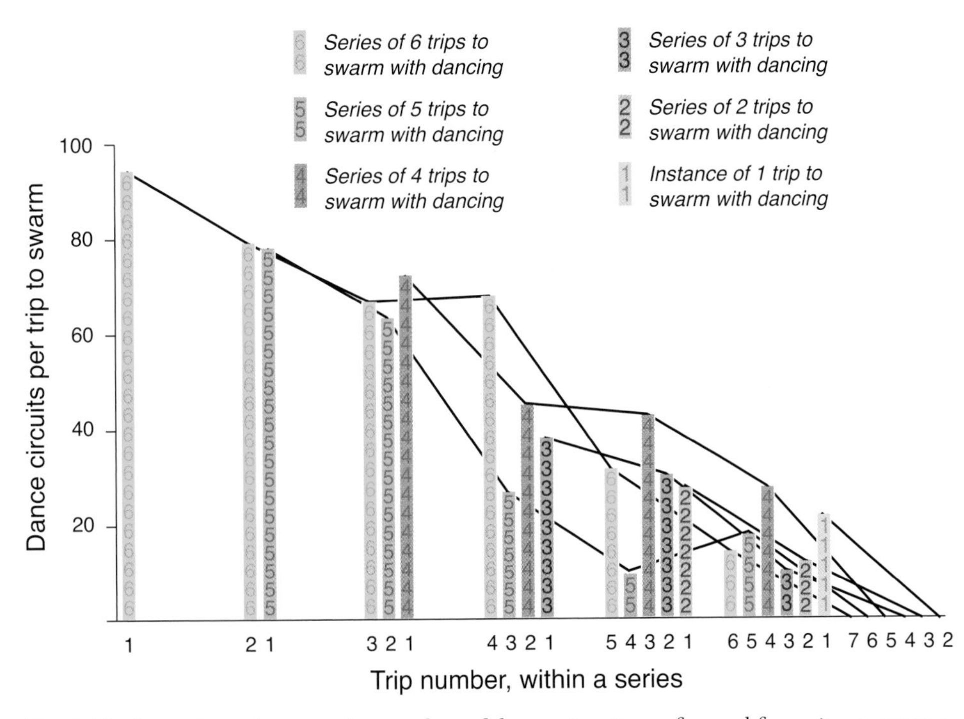

Fig. 6.10 Scout bees decrease the number of dance circuits performed for a site on successive trips back to the swarm. A string of consecutive trips back to the swarm with dancing for a particular site is called a "series"; series vary in length from six trips to one trip. The drop in dance strength per trip (about 15 dance circuits) appears to be a constant, regardless of series length.

site will tend to start her reporting with a smaller number of dance circuits (see fig. 6.5). Because the rate of decay in dance strength per trip back to the swarm is the same for all scouts, a bee from a high-quality site will tend to advertise her site over many consecutive trips back to the swarm (for example, the bee Orange in figure 6.5) and in sum will produce a strong advertisement with many dance circuits, whereas a bee from a medium-quality site will tend to advertise for only a few consecutive trips back to the swarm (for example, the bee Blue-White in figure 6.5) and in sum will produce a weaker advertisement with fewer dance circuits. Consequently, as shown in figure 6.11, a scout bee supporting a superb site, relative to one supporting a poorer site, will be both a longer and "louder"

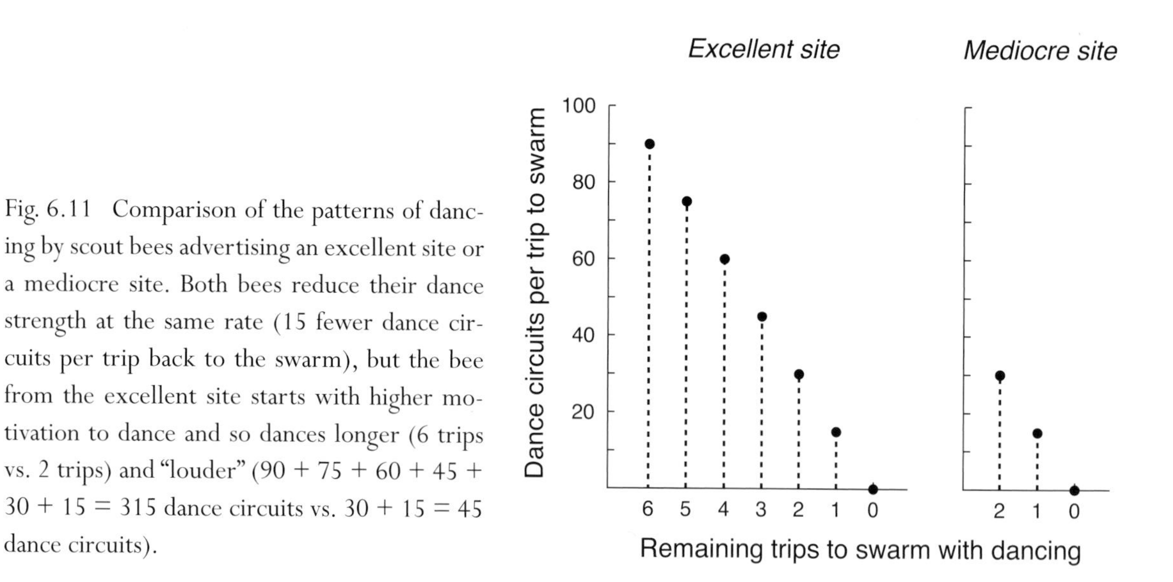

Fig. 6.11 Comparison of the patterns of dancing by scout bees advertising an excellent site or a mediocre site. Both bees reduce their dance strength at the same rate (15 fewer dance circuits per trip back to the swarm), but the bee from the excellent site starts with higher motivation to dance and so dances longer (6 trips vs. 2 trips) and "louder" (90 + 75 + 60 + 45 + 30 + 15 = 315 dance circuits vs. 30 + 15 = 45 dance circuits).

supporter of her site. And as we all know, in any contest for popular support, the side with the most persistent and most zealous supporters is the one most likely to prevail.

It appears, therefore, that a swarm's scout bees do something sharply different from what humans do to reach a full agreement in a debate. Both bees and humans need a group's members to avoid stubbornly supporting their first view, but whereas we humans will usually (and sensibly) give up on a position only after we have learned of a better one, the bees will stop supporting a position automatically. As is shown in figure 6.5 and figure 6.9, after a shorter or longer time, each scout bee becomes silent and leaves the rest of the debate to a new set of bees. Figure 6.7 shows how this regular turnover in which scouts are dancing can help a swarm's scouts quickly reach an agreement, for in this schematic depiction of consensus-building on a swarm all of the bees that were active dancers at 10:00 a.m. have retired by 1:00 p.m., and all those that were active dancers at 1:00 p.m. have retired by 4:00 p.m.

There is, however, one important case in which human group decision making operates in a manner similar to that of honeybee swarm house-hunting. It is how scientists conduct their social decision making on scientific theories. Many have noted that new and better ideas succeed in scientific debates through attrition, that is, by one generation of scientists retiring from their field and eventually dying off. But before this generation drops from the debate, the next generation of scientists will have listened carefully to the various arguments made by their predecessors, been persuaded by the most compelling claims on the truth, and adopted the new theory. Thus the support for a new and better theory (e.g., the sun-centered theory of Copernicus and Galileo) grows while it fades for an older and poorer one (e.g., the earth-centered theory of Ptolemy). The most often quoted statement describing this social process is by Max Planck: "A new scientific truth does not triumph by convincing its opponents and making them see the light, but rather because its opponents eventually die, and a new generation grows up that is familiar with it." One difference between aged scientists and aged scouts, though, is that the people tend to drop out of the debate reluctantly, sometimes not until death, whereas the bees do so automatically. I cannot help but wonder whether science would progress more rapidly if, in this regard, people behaved a bit more like bees.

7

INITIATING THE MOVE TO NEW HOME

And so doth this soft shivering passe
as a watch-worde from one to an other,
untill it come to the inmost Bees:
wherby is caused a great hollownes
in the pomgranat.
When you see them do thus,
then may you bid them farewel:
for presentlie they begin to unknit,
and to be gone.
—*Charles Butler,* The Feminine Monarchie, *1609*

Anyone who has the immense good fortune of watching a honeybee colony cast a swarm will be treated to many astonishing displays of animal behavior. First there is the feverish rush of thousands of bees out of the hive and up into the sky. Minutes later, the cloud of swirling, swarming bees mysteriously condenses into a tight crowd hanging from a tree branch, where for several hours or several days nearly all the bees sit quietly, almost motionless. Only the swarm's scouts remain active, flying to and from the swarm cluster and performing their eye-catching dances on its surface to advertise candidate nest sites. Next, after one of these sites becomes the unanimous choice of all the dancing bees, comes the most wondrous sight of all: suddenly, in about 60 seconds, the entire swarm cluster disintegrates and takes flight, filling the air with the roar of thousands of airborne bees (fig. 7.1). This flying mob immediately begins moving off in the direction of

Fig. 7.1 The author, on Appledore Island, watching a swarm launch into flight from the vertical board that he uses as a swarm mount. The two feeder bottles on the mount provide sugar syrup to keep the swarm well fed.

its chosen home and in another minute or two it will have vanished. As Charles Butler expressed it so nicely back in 1609, you may now "bid them farewel."

In this chapter we will look at how an entire swarm of bees manages to leave its bivouac site together and at the right time. With few exceptions, a swarm's tightly synchronized takeoff occurs only after its scouts have finished their job of choosing the new dwelling place. This means that as we review the mechanisms of social coordination during swarm departure, we will be seeing how a swarm maintains its coherence as it switches its mission from making a decision to implementing a decision. Probably it will be no surprise to learn that the scout bees are the rabble-rousers who initiate a swarm's journey to its new home, thereby extending their leading role in our story into the chapters on how swarms take action. But what will be surprising are the nifty signals the scouts use to animate their drowsy swarm-mates and the way the scouts know when the time has come to initiate their swarm's journey. Indeed, until recently, these were deep mysteries about the inner workings of honeybee swarms.

Preflight Warm-Up

In the spring of 1980, Bernd Heinrich, a gifted insect physiologist at the University of California at Berkeley (now at the University of Vermont), turned his attention to the mechanisms of temperature regulation in honeybee swarms. Over the previous 20 years Heinrich had pioneered the study of temperature control in insects, so he began his study of honeybee swarms with a great deal of background knowledge. He knew that two previous studies had reported that the temperature inside a swarm cluster, like that inside a beehive, can be kept by the bees at about 35°C (95°F), nearly the same as the core body temperature of a human. He also knew that an individual worker honeybee can produce heat by shivering—isometrically contracting the two sets of flight muscles in her thorax—and that her flight muscles must be warmed to at least 35°C (95°F) to produce a sufficiently high wing-beat frequency (nearly 250 beats per second!) to generate the lift needed for bee flight. Furthermore, Heinrich knew that before leaving the parental nest, swarming honeybees stuff themselves with honey, so that a swarm starts out with a sizable but finite supply of fuel for warming itself, for powering the scout bees' flights to and from the swarm, and for building the first beeswax combs in the swarm's new home. What he did not know were the exact pattern of temperatures inside a swarm cluster, how the bees control these temperatures, and how they manage their energy supply. Being a hobby beekeeper and curious about bees, Heinrich worked with the police and fire departments of Walnut Creek, California, as well as the Ecohouse Swarm Hotline of Walnut Creek, to collect 14 natural swarms in the San Francisco Bay Area during May and June. Back at his laboratory on the UC-Berkeley campus, he studied these swarms using various scientific tools, including tiny electronic thermometers (thermocouple probes) and a special cylindrical chamber made of Plexiglas (respirometry vessel) in which he could place a swarm to measure its metabolic rate at various ambient temperatures.

Fig. 7.2 Top: The mantle bees in a swarm when the ambient temperature is 28°C (82°F). Bottom: Same bees at 13°C (55°F). When cool, the mantle bees huddle more tightly and reduce the mantle's porosity.

Heinrich discovered many marvelous things about temperature regulation in honeybee swarms, all of which are key to understanding how a swarm prepares to fly to its new home. First, he found that a swarm does indeed precisely control the temperature of the cluster's core so that it stays at 34–36°C (93–97°F) regardless of the ambient temperature. He also found that a swarm allows the temperature of the cluster's mantle (outer layer) to vary with the ambient temperature, but that it keeps the mantle temperature above 17°C (63°F) even if the ambient temperature falls to freezing (0°C or 32°F). This means that the outermost bees, which are the coolest, keep themselves warm enough to stay active on the swarm. If they were to cool below 15°C (59°F) they would enter "chill torpor" and easily fall from the swarm. They would also be too cold to warm themselves back up by shivering.

When Heinrich looked at how the bees achieve their characteristic pattern of temperatures in a swarm, he found that they do so without expending much of their on-board energy supply, that is, the honey in their stomachs. At air temperatures above about 10°C (50°F), the *resting* metabolism of a swarm—the metabolism that occurs when the flight muscles of the swarm bees are not being activated—provides more than enough heat to keep a swarm's core at 35°C and its mantle above 17°C. Indeed, at high ambient temperatures (above about 20°C or 76°F), the resting metabolism produces so much heat that both the mantle bees and the core bees spread themselves out, creating ventilation channels to release excess heat from the core. But when the ambient temperature falls below 17°C, and the mantle bees start to feel too cool, they crowd inward, causing the swarm cluster to shrink, its porosity to decrease, and its heat loss to diminish (figs. 7.2 and 7.3). In this way the mantle bees skillfully trap inside the swarm cluster the metabolic heat generated by the thousands of resting, immobile bees, and they also keep themselves sufficiently warm. It is only when the air temperature falls below about 10°C (50°F) that the mantle bees must take the extra step of raising their metabolic rate by shivering.

Thus Heinrich discovered that the bees in a honeybee swarm have an effective means of conserving their energy reserves. The mantle bees, those most exposed to low temperatures, minimize their need for active metabolism by doing two things when the air becomes cool: (1) letting their body temperatures drop

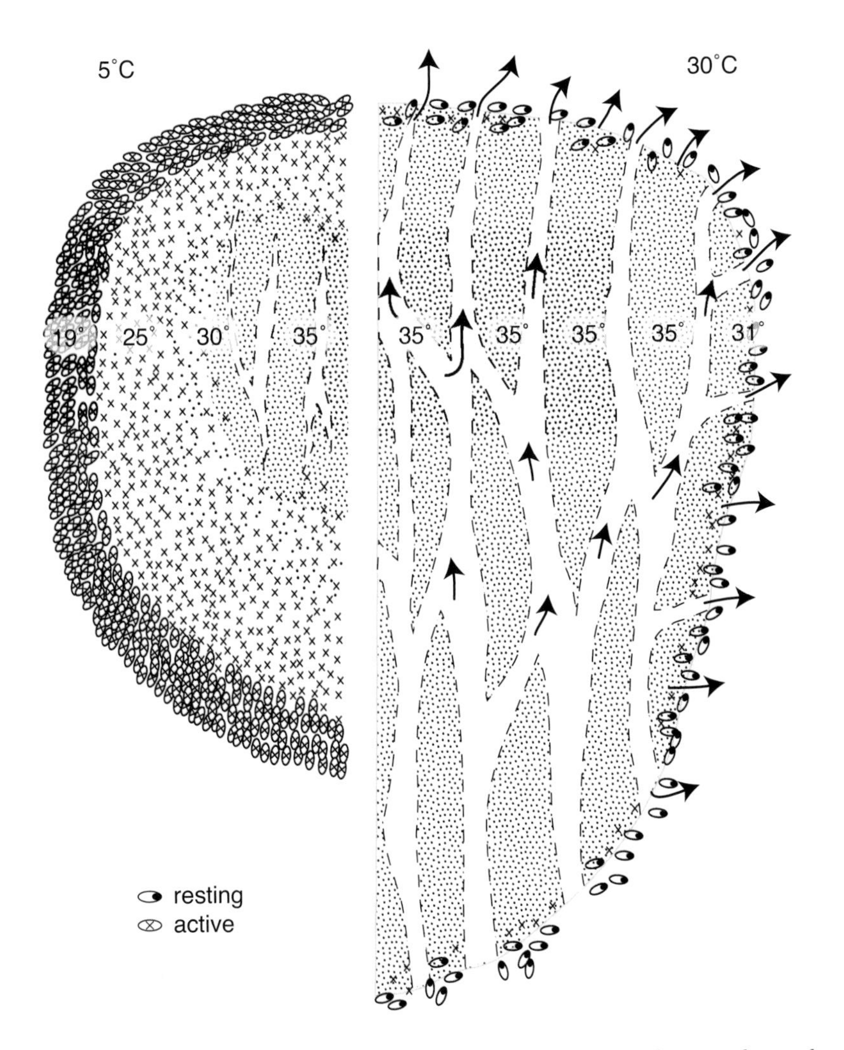

Fig. 7.3 Diagram summarizing how the bees in a swarm cluster achieve thermoregulation at a low (left) and a high (right) ambient temperature. Shown are the positions of bees, channels for ventilation, losses of heat (arrows) and areas of active metabolism (crosses) and resting metabolism (dots).

to just above the chill-torpor temperature rather than working to maintain a higher body temperature, and (2) keeping their body temperatures above the chill-torpor temperature mainly by huddling rather than shivering. Of course, these energy conservation measures mean that most of the time the outermost bees in a swarm are too cold to fly, something that is easily demonstrated by skimming a spoonful of mantle bees from a swarm and shaking them into the air. The bees tumble to the ground rather than fly away. So before a swarm can take off to fly to its new home, the cool bees in the mantle must warm their flight muscles to the flight-ready temperature of 35°C. And not just in theory! When Heinrich made continuous recordings of the temperatures at various locations in a swarm cluster from when the bees settled to when they departed, he found that during the last hour or so before takeoff, the temperature in the mantle did indeed rise to match the 35°C of the core.

In June 2002, some 20 years after Bernd Heinrich published his insightful report on "The Mechanisms and Energetics of Honeybee Swarm Temperature Regulation," I traveled to Germany to look more closely at the preflight warm-up of swarm bees. Shortly before, I had the great good fortune of receiving a Research Prize from the Alexander von Humboldt Foundation, which gave me the wherewithal to conduct research projects in Germany. I was hosted by my teacher and friend, Bert Hölldobler, who had become the director of the Institute for Behavioral Physiology and Sociobiology at the University of Würzburg. This institute includes a laboratory devoted to research on honeybees. It was started by Martin Lindauer when he was professor of zoology at Würzburg (1973–1987), and is now directed by another good friend, Jürgen Tautz. Jürgen is highly skilled at studying the sensory abilities of insects, and his laboratory is stocked with much state-of-the-art scientific equipment for probing the workings of nature. On this trip I was keen to collaborate with Jürgen to use one particularly powerful instrument: an infrared video camera. With it, one can measure the temperatures of many objects (such as bees) simultaneously and without disturbing them. Also in Jürgen's laboratory were two superb graduate students, Marco Kleinhenz and Brigitte Bujok, both experts in using the video camera and the computer software that converts the camera's images into accurate temperature readings. The goal of our four-person team was simple: explore how the mantle bees in a honeybee swarm warm their flight muscles prior to takeoff.

The plan of using video thermography to see how the outermost bees prepare for takeoff worked nicely. Over a two-week period, we recorded the temperatures of mantle bees within a 10 x 10 centimeter (4 x 4 inch) area on two

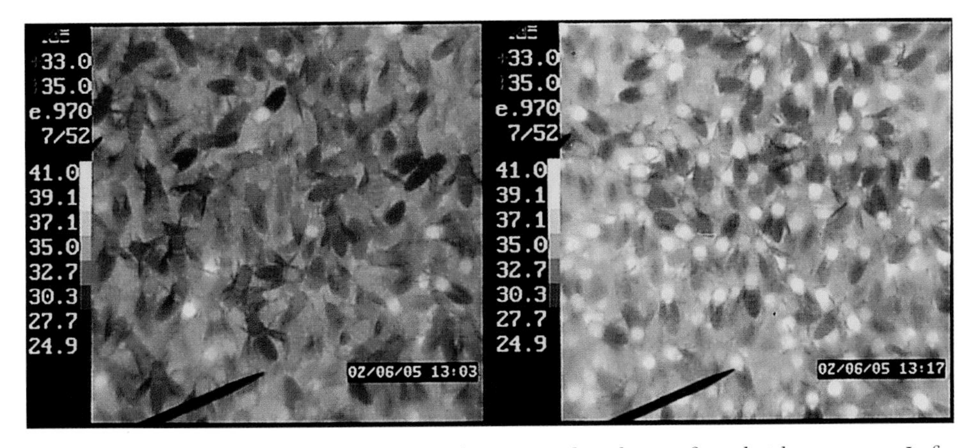

Fig. 7.4 Bees on the surface of a swarm when viewed with an infrared video camera. Left: Image made 15 minutes before takeoff. Right: Image made only 1 minute before takeoff. On the left in each image is a scale bar indicating the temperature (in °C) represented by each shade of gray.

swarms, starting when each swarm formed its cluster and continuing until it launched into flight. Both swarms showed the familiar set of events shortly before takeoff: the scouts became unanimous in their dancing and the nonscouts began to move excitedly. Both swarms also revealed something new in the images recorded with the infrared video camera (fig. 7.4): the thoraces of *all* the bees on the swarm's surface began to glow with unusual warmth just moments before the swarm's explosive takeoff.

The finding that most captured our attention was the way that the percentage of bees with a thoracic temperature of at least 35°C rose exponentially over the final half hour before takeoff. As is shown in figure 7.5, for the first 20 minutes the percentage of the surface-layer bees with thoraces warmed sufficiently for flight rose slowly and remained below 20 percent. Then, starting about 10 minutes before takeoff, the percentage of hot bees began to rise faster and faster. Soon 100 percent of the surface-layer bees had a thoracic temperature of at least 35°C, and at exactly this moment the swarm bees took wing. We are confident that at the start of a swarm's takeoff all of the bees in the cluster, not just the outermost ones, are hot enough for rapid flight. After all, Heinrich's work had shown that the bees in a swarm cluster's core are warm enough for flight at all times.

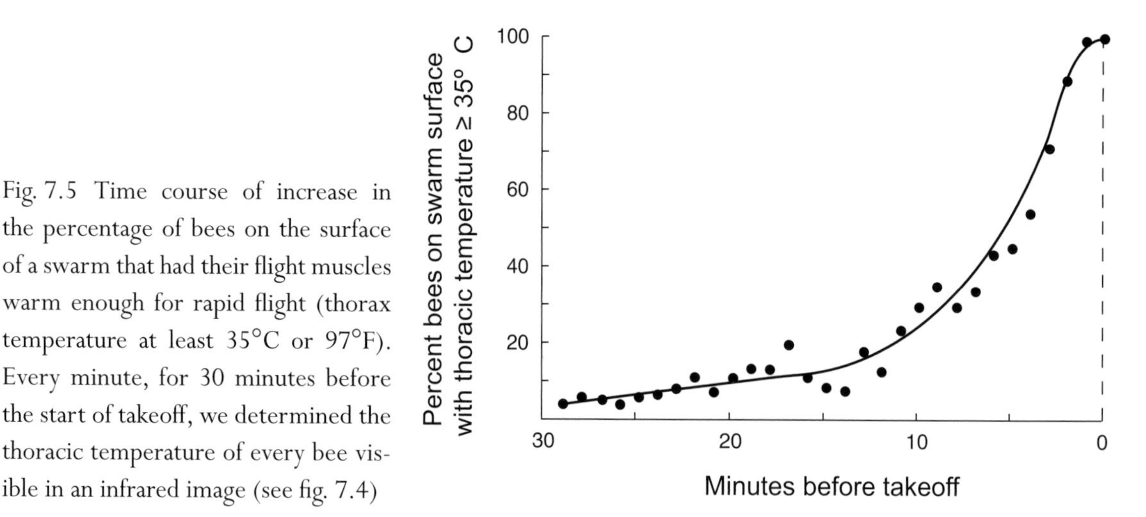

Fig. 7.5 Time course of increase in the percentage of bees on the surface of a swarm that had their flight muscles warm enough for rapid flight (thorax temperature at least 35°C or 97°F). Every minute, for 30 minutes before the start of takeoff, we determined the thoracic temperature of every bee visible in an infrared image (see fig. 7.4)

Also, the images from our infrared video camera showed that as our two swarms approached the moment of takeoff, the interior bees began to shine brightly before the surface bees did, looking like hot coals glowing beneath a layer of cool ashes. There is also the fact that immediately after both takeoffs had started—that is, when the outermost bees had taken flight—the inner bees began to take flight. Indeed, it is because there is so little delay between outer and inner bees taking flight that a swarm cluster needs only about 60 seconds to disintegrate.

What stimulated the mantle bees to warm up, and how was it that each swarm's takeoff began just seconds after all of the surface-layer bees had warmed their flight muscles to at least 35°C? In other words, what stimulated the bees to prime themselves for flight, and what finally triggered them to take flight? We will now probe these two mysteries.

Piping Hot Bees

If you listen closely to a swarm, by carefully placing an ear beside the massed bees, you will hear pulses of a distinctive, high-pitched piping sound starting about an hour before the swarm flies off to its new home. Each sound pulse lasts about a second, and because its pitch sweeps upward, it resembles the rising engine whine of a Formula One race car making a quick acceleration. At first one hears these shrill piping sounds only occasionally, because just one bee at a time is producing them, but over the last half hour before takeoff more and more bees start piping and the pulsing hum radiating from the swarm rises to a crescendo. When it does, the swarm cluster breaks up and all the bees take wing. Could this high-pitched piping be a signal from the scout bees to their quiet swarm-mates with the message, "Ladies, warm your flight muscles!"?

To begin to explore this possibility, I wanted to identify which bees in the swarm cluster were producing the high-pitched piping sounds. Actually, this was a long-standing goal. I had first heard these mysterious sounds way back when I began studying swarms as a graduate student in the 1970s, but I never could pinpoint which particular bees, among the thousands in a swarm, were their source. Finding the piping bees was especially difficult because the pulsing hum seemed to emanate from inside the swarm cluster, thus from bees out of sight. The piping bees also stymied Martin Lindauer in the 1950s, for he wrote, "Now a hundredfold high humming could be heard at the cluster, but I could not definitely find out whether this comes from the buzz-runners or from other bees." (The "buzz-runners" mentioned by Lindauer will be discussed later in this chapter.)

The discovery of who does the piping came serendipitously in the summer of 1999. It started with a chance observation I made at my camp beside Ox Cove, in easternmost Maine, the wonderfully secluded site to which I had retreated to figure out how the dissent among dancing scout bees expires (see chapter 6). I can still remember as if it were yesterday witnessing for the first time a worker piping on a swarm. I had set up a swarm outside my cabin, labeled with paint dots the first few dancers (scouts) on the swarm, and was watching steadily my little band of brightly painted bees, recording their behaviors. On August 2, at 10:48 a.m., just five minutes before the swarm flew away, my attention was drawn to the scout bee Blue, who did something unexpected on the swarm's surface: she ran excitedly over other bees for a few seconds, then paused for about a second, pressed her thorax against a stationary bee, and then ran on, repeating the sequence of run-pause-press six times before she burrowed into the cluster and disappeared (fig. 7.6). I noticed that each time my bee Blue paused and grabbed another bee, she drew her wings tightly together over her abdomen and then her wings seemed to vibrate slightly.

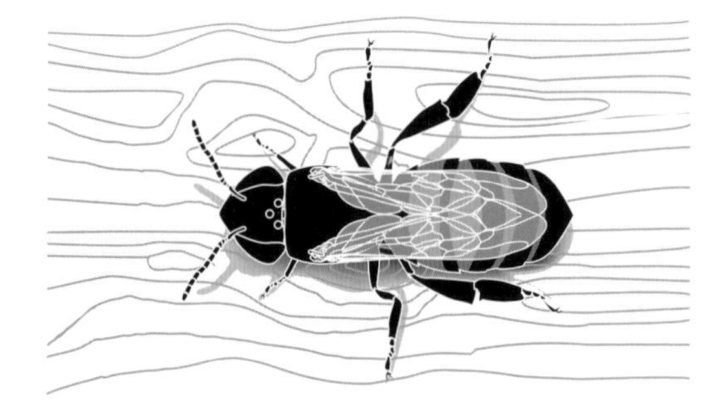

Fig. 7.6 A worker bee producing the piping signal. During a pause from running over bees in the swarm cluster, she presses her thorax to the substrate, pulls her wings together tightly over her abdomen, arches her abdomen, and activates her wing muscles to produce a vibration in the substrate. Although the substrate shown here is a wooden surface, almost always the substrate is another bee.

Was Blue producing the piping sound? I could hear the sound, but with just my "naked" ears I could not be sure the sound came from her. So that afternoon, I drove to Morgan's Garage in the nearby village of Pembroke and bought a 3-foot length of rubber vacuum hose about 6 millimeters (one-quarter of an inch) in diameter, a size that would fit snugly in my ear. This simple sound tube would enable me to localize the source of sounds coming from my swarms, for it would function like a primitive stethoscope, conducting to my ear only sounds produced near its open end. A few days later, when I watched a second swarm and used my rubber hose to listen in on another painted scout bee doing the run-pause-press maneuver, I was thrilled to hear the perky piping sound.

I was fascinated by the sights and sounds of the piping worker bees, and was keen to describe their signal in detail and to test the idea that they are alerting the quiescent members of the swarm to warm their flight muscles for takeoff. This would require a sophisticated sound analysis combined with careful observations and experiments. Fortunately, Jürgen Tautz was easily persuaded to join in the venture, and in August 2000, he joined me at Cornell, bringing with him from Germany the miniature microphones and digital audio and video equipment that we would need for the project. Soon we had a swarm set up in a quiet spot at my laboratory, with the swarm clustered on one side of a vertical board so that we could easily monitor everything that happened on the swarm's surface. Inside the swarm we mounted two microphones and several temperature probes, and directly in front of the swarm we positioned a video camera that recorded both the bees' sounds from the swarm's interior and the bees' actions on its surface. With numerous microphone and thermometer wires leading from it, a video camera continuously recording its activity, and two biologists hovering over it, our swarm looked rather like a patient in an intensive care unit.

Because I now had a search image for a piping bee—one dashing over the swarm's surface but pausing frequently to seize a motionless swarm-mate—I was able to spot pipers at a glance when we started hearing their shrill piping sounds. From our video recordings, Jürgen and I quickly confirmed my previous observation that the piping bees are exceptionally excited scout bees. The bees made this crystal clear by switching between worker piping and waggle dancing while scrambling over the surface of the swarm (fig. 7.7). This mixing of signals became especially noticeable during the last half hour before takeoff, when the piping grew strongest. (How the scouts know when to start piping will be explained later in this chapter.) We saw then that after a scout bee had finished a bout of waggle dancing, she was very likely to start producing a string of piping signals.

From our audio recordings of pipers sounding off near one of the microphones, we learned that each pipe is a single pulse of sound that lasts about one second and is composed of a fundamental frequency of 200 to 250 hertz (cycles per second) plus many harmonics—multiples of the fundamental frequency—in the range of 400 to 2,000 hertz (fig. 7.8). It is these high-frequency harmonics that make each pipe sound so shrill. That the fundamental frequency of the piping sound matches the wing-beat frequency of a flying bee is strong evidence that a bee produces this sound by activating the flight muscles in her thorax to create strong vibrations in her body. Probably most of this vibration energy passes as a sharp blast into the bee that the piper has grabbed hold of and pressed against, but some passes into the surrounding air creating the sounds that humans can hear while eavesdropping on a swarm. Jürgen and I also learned from our sound recordings that the upward sweep in the pitch of each pipe is produced by a shift

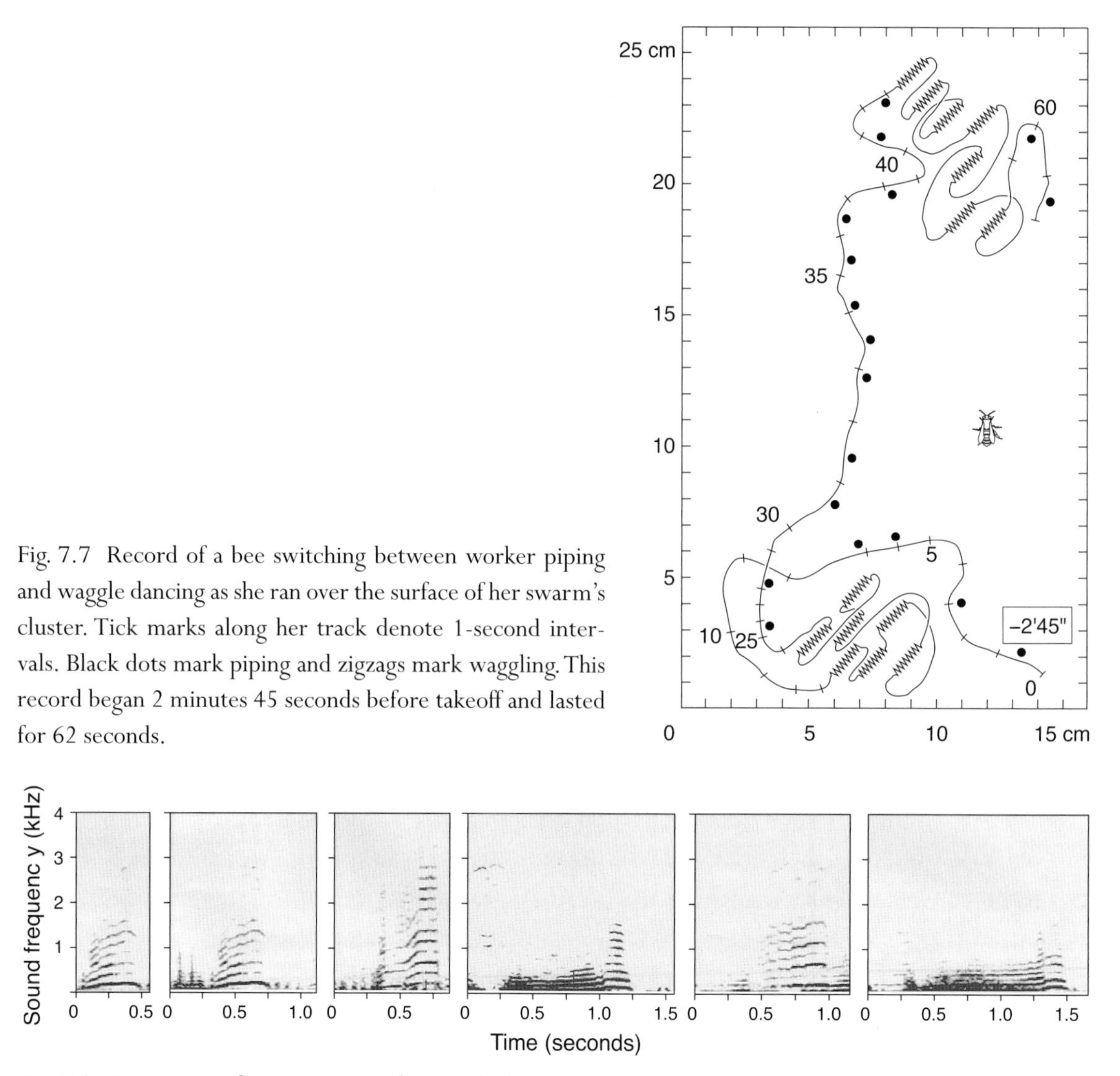

Fig. 7.7 Record of a bee switching between worker piping and waggle dancing as she ran over the surface of her swarm's cluster. Tick marks along her track denote 1-second intervals. Black dots mark piping and zigzags mark waggling. This record began 2 minutes 45 seconds before takeoff and lasted for 62 seconds.

Fig. 7.8 Sonograms of six piping signals recorded from workers in a swarm shortly before takeoff. The units shown on the vertical axis are kilohertz, or thousands of cycles per second.

in the fundamental frequency from 200 to about 250 hertz and an increase in the amount of sound energy in the high-frequency harmonics. Probably a piper creates these changes by pulling her wings together, thereby stiffening her thorax and raising its resonant frequency.

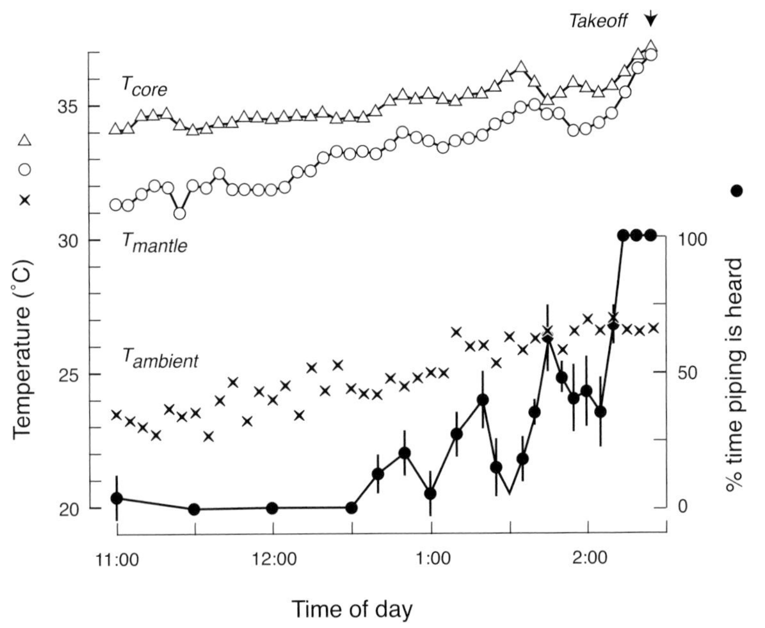

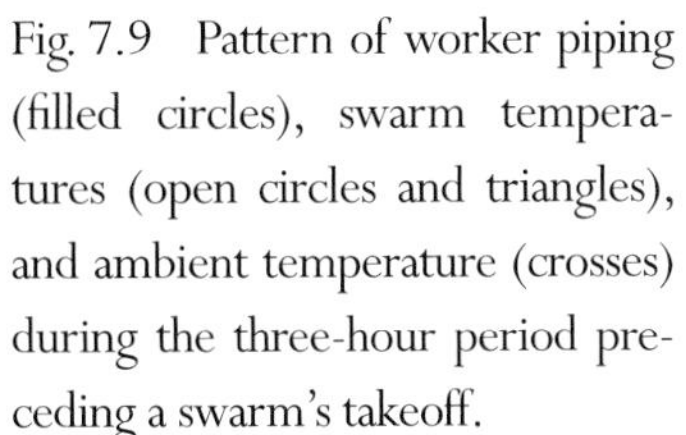
Fig. 7.9 Pattern of worker piping (filled circles), swarm temperatures (open circles and triangles), and ambient temperature (crosses) during the three-hour period preceding a swarm's takeoff.

At this point Jürgen and I wanted to test the hypothesis that the function of worker piping in swarms is to stimulate bees to prepare for takeoff. Our first step was a check that worker piping really occurs only in the last hour or so before takeoff, when the bees in a swarm are making their flight preparations. We did this by measuring simultaneously the level of piping in a swarm and the temperatures in the swarm's core and mantle for many hours prior to takeoff. Figure 7.9 shows an example of the patterns in piping and warming that we found. Three hours before takeoff (11:30 a.m.), when the ambient temperature was 23°C (73°F) and the swarm's core and mantle temperatures were 34° and 31°C (93° and 87°F), we heard no piping. Then, about 90 minutes before takeoff, we began to hear piping but only intermittently. Finally, during the half hour before takeoff, the sound of the piping bees was continuous and loud, for by then multiple bees were piping simultaneously. At the same time, the mantle temperature began rising, and just when the temperature throughout the clustered swarm reached 37°C (99°F) it launched into flight.

The finding that worker piping coincides perfectly with swarm warming—both

probe (fig. 7.11). Both cages soon became filled with mantle bees. We prevented the scouts from contacting—hence sending piping signals to—the bees inside one of the cages by closing it with a screen cover when we started hearing bees piping. Simultaneously, we treated the bees inside the other cage in exactly the same way, except that we "closed" their cage with a cover that had a large opening through which pipers could pass. Given our hypothesis that piping bees stimulate bees to warm up for takeoff, we predicted that the mantle bees in the closed cage would not warm themselves to a flight-ready temperature at the time of takeoff, whereas those in the open cage would do so. This is precisely what we found. The open-cage bees showed the usual pattern of a dramatic rise to approximately 35°C (95°F) in the final minutes before takeoff, but the closed-cage bees did not (fig. 7.11). For fun, at the end of each trial, after all the uncaged and open-cage swarm bees had departed, we removed the cover of the closed cage and prodded the bees inside, who were eerily calm. All tumbled to the ground, too cold to fly. Without a doubt, these caged bees had missed the scout bees' persistent alerts to warm up for takeoff.

Boisterous Buzz-Runners

The investigation of the piping signal solved the mystery of how the scout bees prime their swarm-mates for flying to their new home, but it left unsolved the puzzle of what finally triggers the highly synchronized, virtually explosive, takeoff of the thousands of bees in a swarm. A strong possibility was an eye-catching behavior that Martin Lindauer, who first described it, named the *Schwirrlauf*. English-speaking bee biologists call this behavior the buzz-run. It is well named in both German and English because a buzz-running bee runs across the swarm cluster, turning this way and that, usually with her outspread wings whirring and buzzing noisily. Sometimes she is dashing over the backs of the immobile bees and other times she is bulldozing between them (fig. 7.12). Lindauer reported that buzz-runners are prominent on the swarm cluster in the final few minutes before takeoff starts, and he suggested that by barging and boring through the cluster, the buzz-runners disperse the interconnected bees and so initiate their concerted takeoff. This was an enchanting hypothesis, but it remained to be tested. And

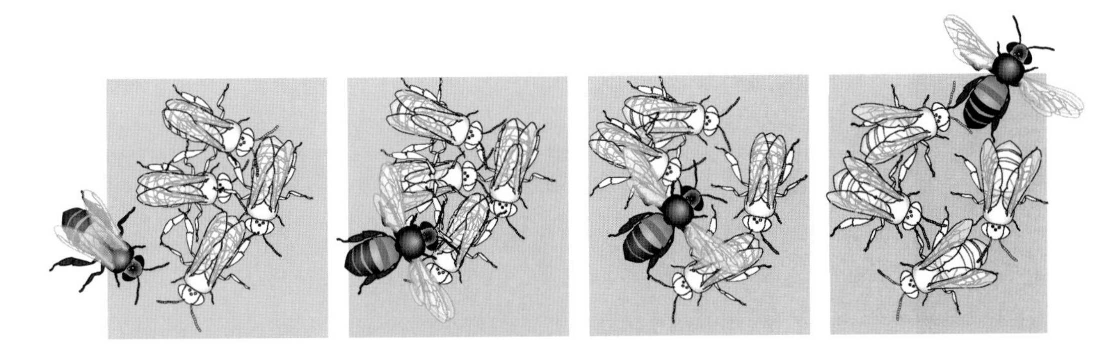

Fig. 7.12 A worker bee performing the buzz-run through a small group of lethargic bees. Panel 1: The buzz-runner runs toward the knot of bees. Panel 2: One second later, the buzz-runner has spread her wings and is buzzing them as she makes contact with the cluster. Panel 3: One second after making contact, the buzz-runner is pushing through the cluster, still buzzing her wings. Panel 4: She has broken contact with the bees but continues buzzing her wings as she runs on. Based on frames of a video recording.

even if it proved correct, many questions would remain about the hyperactive buzz-runners. What is the interplay between worker piping and buzz-running as a swarm prepares for and then takes to flight? Which bees in a swarm perform buzz-runs? And how do buzz-runners know when to produce their rough signal?

In tackling these questions, I was joined by Clare Rittschof, a Cornell undergraduate student who turned out to be a born researcher. We began our investigation of The Case of the Buzz-Runner Bees in May 2005, as soon as Clare had finished the final exams in her spring semester classes. We started with a stakeout for buzz-runners on swarms to find out when they conduct their activities. To do this, we mounted a swarm of bees on one side of a vertical wooden board and video recorded the activities of the bees within a 10 x 15 centimeter (4 x 6 inch) region of the swarm's surface. We started each round of surveillance when the swarm bees began to produce piping signals and ended it when the bees flew off to their chosen home. Clare would play back the recordings in slow motion, scanning them for bees running erratically over the swarm's surface. My studies of worker piping led us to expect that some of the running bees would be pipers, and Lindauer's report indicated that others would be buzz-runners. To know whether or not any given runner was a piper, we followed each one for several

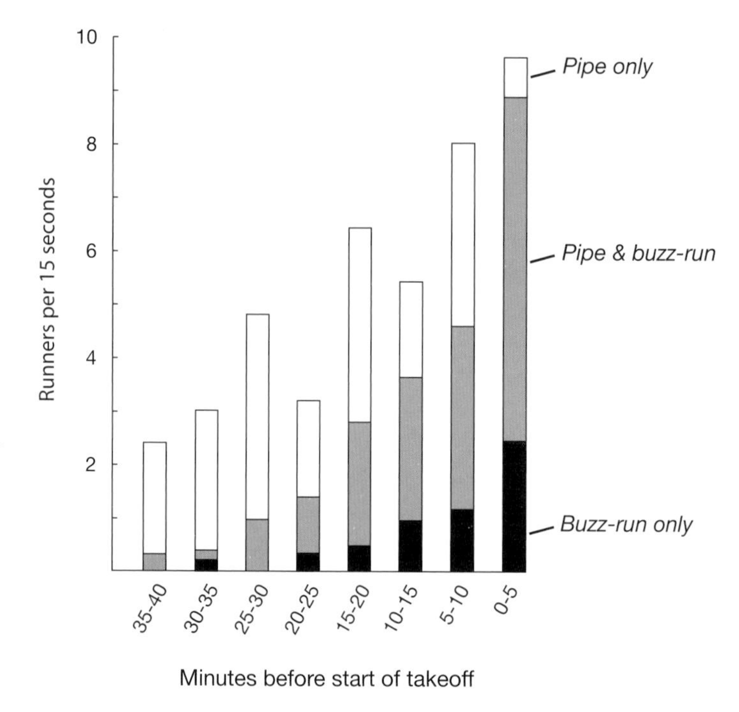

Fig. 7.13 Record of the increase in number of scouts running across a 10 x 15 centimeter (4 x 6 inch) region of a swarm's surface in 15 seconds during the 40-minute period preceding swarm takeoff. Also shown is the changing pattern of signal production by the running bees, all of which produced pipes or buzz-runs, or both.

seconds with a small microphone (to pick up any piping sounds she might be producing) and added this audio information to the video recording. Buzz-runners were easily identified among the running bees by the conspicuous buzzing of their outspread wings.

Clare's painstaking examinations of our surveillance records yielded two important findings. First, she saw that more and more runners went into action during the final hour preceding takeoff so that just before its takeoff a swarm teems with bees dashing over and through the cluster. Second, and more remarkably, she saw that *all* of the runners produced audible signals: pipes or buzzes, or both. At first, the running bees produced just piping signals. But little by little, they started to combine piping with buzz-running—ramming into other bees and revving their wings—and during the final five minutes before takeoff more than 80 percent of the running bees produced buzz-runs (fig. 7.13). This told us that the buzz-runners are the same bees as the pipers, who we already knew to be scout bees. Thus we learned that the scout bees give both the piping signal to prime the swarm for takeoff and the buzz-run signal to trigger (evidently) the takeoff.

What is the evidence that the buzz-run signal stimulates bees to take flight? First, there is the fact that the buzz-run is an ephemeral signal that is seen in just one context: when idle bees are being stimulated to take flight. So we see buzz-runners briefly just before a swarm pours forth from its hive (as described in chapter 2) and again just before it takes wing from its bivouac site. There is also the fact that buzz-run production rises to a crescendo moments before swarm takeoff, suggesting that the former causes the latter. And perhaps most telling, there is what Clare found when she reviewed the video recordings of many episodes of a buzz-runner blasting through a knot of lethargic bees. She found that they were more dispersed and more active after they had experienced the buzz-runner's forceful persuasion than before.

One feature of a buzz-runner's behavior that should be noted is that sometimes she will launch herself into flight, fly around the swarm cluster for a few seconds, and then land back on the cluster and resume her buzz-running. The phenomenon of buzz-runners taking flight is important because it points to the evolutionary origins of their lively signaling behavior. Almost certainly, the buzz-run signal is a ritualized form of a bee's takeoff behavior, which consists of a bee spreading her wings, starting to buzz them, pushing clear of other bees if need be, and finally taking to the air.

"Ritualization" is the name biologists have given to the process whereby some incidental action of an animal becomes modified over evolutionary time into an intentional signal. Usually the incidental action is a by-product of an activity performed in one particular context, so the animal's action is a reliable indicator of this context. The buzz-run illustrates this idea nicely: when a bee is about to take flight, she inevitably buzzes her wings, so wing buzzing by a bee is a reliable indicator to others that she is about to take flight. The next step in a signal's evolution is for the receivers to detect it and use the information it provides to improve their decision making. If the receivers' improved decision making also benefits the senders, then the senders will benefit by making the signal more conspicuous and so more easily detected by the receivers.

In the early stages of the evolution of the buzz-run signal, the quiescent bees in a swarm probably improved their decision making about when to take flight by responding to the wing buzzing of other bees taking off. The improved decision making by the quiet bees probably produced better-coordinated takeoffs, which also benefited the active bees, so natural selection favored modifications of the wing buzzing by the active bees to make it more conspicuous to the quiet bees. Given the present-day form of the buzz-run, it appears that these modifications include exaggerating the wing buzzing (starting it long before the moment the buzz-runner takes flight) and adding to it the actions of running and ramming. I think the buzz-run shows nicely how sometimes we can glimpse the evolutionary origins of the marvelous signals that bind bee to bee to bee in a swarm.

One final question regarding the buzz-run is why did honeybee swarms evolve this signaling system? In other words, why should the scout bees send everyone else in the swarm a signal of when to launch into flight? I suggest that this signaling system evolved because it is only the peripatetic scout bees that can sense when all the bees in the swarm cluster are ready for departure, and the buzz-run signal enables the scouts to share this critical information with their swarm-mates. As we have seen, for all the bees in a swarm to launch into flight together each bee must have her thorax warmed to at least 35°C (97°F). But how can *all* the bees in a swarm know when they've *all* become hot enough? One way would be to have some bees travel across the swarm cluster, with each one measuring the temperatures of her swarm-mates along the way, and then sounding a departure alarm when her canvassing tells her that the required warmth has been achieved by all. I suspect that this is how it works on swarms, for we now know that the scout bees move quickly throughout the swarm cluster, with each scout pausing every few seconds to press her thorax against another bee and produce the piping signal. Perhaps each time a scout presses against another bee she also senses her temperature. And we now know that it is the scout bees that strongly produce the buzz-run signal in the final few minutes before takeoff, when all the bees have the high body temperature required for departure.

If the hypothesis of scout bees as mobile temperature sensors, information integrators, and group activators proves correct, then the mechanisms mediating the initiation of takeoffs by honeybee swarms present us with an intriguing system of behavioral control within a large group. It is one in which a small minority of individuals actively poll the group to collect information about its global state and then, when the group reaches a critical state, these individuals produce a signal that triggers an appropriate action by the whole group. The governance of a honeybee swarm is proving ever more extraordinary.

Consensus or Quorum?

We know that a swarm starts to switch from *making* a decision about its future home to *implementing* this decision when its scout bees start to produce the piping signals that inform the nonscout bees that the time has come to warm up their flight muscles. So far, so good. But how do the scout bees know when to start producing their piping signals? Given the striking way that the dances on a swarm come to represent one site and then the swarm moves to this site, it is tempting to think that the scouts use the appearance of dancer consensus to know when to start piping, rather like Quakers discuss and wait to find common ground and then, recognizing they have reached a "sense of the Meeting," know when to take action. By this hypothesis, a scout "votes" in favor of a site by dancing for it, the scouts act and interact (as we have seen in chapter 6) so that gradually their votes come into agreement in favor of a superior site, and somehow the voting pattern of the scouts is steadily monitored so that they know when they've reached an agreement and can start acting on their decision.

There were, however, two facts that cast doubt on this appealing hypothesis. First, neither Lindauer nor I nor anyone else had seen any sign of scout bees polling their fellow dancers, something that surely they would need to do to sense a consensus. Second, Lindauer had seen two out of the 19 swarms that he studied launch into flight without a consensus among the dancers, that is, when there were two strong coalitions of dancers advertising two distinct sites (e.g., his Balcony swarm, see fig. 4.4). Were these cases of takeoff with dancer disagreement bizarre anomalies that should be ignored, or were they valuable clues that should be heeded?

I chose to heed them, and did so in a series of collaborative studies with my good friend and former student Kirk Visscher, who shares my passion for figuring out how honeybee colonies work. I first met Kirk in the fall of 1976, when he enrolled in the Biology of Social Insects class I was teaching at Harvard. We clicked right away. Here was someone who was already extremely knowledgeable about honeybees from years of beekeeping with his father, someone with a powerful intellect, someone with a modest sense of self, a quick smile, and a love of biology. I was to learn later that Kirk is also a fabulous gadgeteer, expert statistician, and computer whiz—all things that I'm not. Today Kirk is on the faculty of the University of California at Riverside.

Even though Kirk and I now live on opposite sides of the North American continent, we teamed up because both of us had long wondered whether the scouts on a honeybee swarm know when to start their piping by sensing a *quorum* (sufficient number of scouts) *at one of the nest sites* rather than by sensing a *consensus* (agreement of dancing scouts) *at the swarm cluster*. By the quorum-sensing hypothesis, a scout "votes" for a site by spending time at it, the number of scouts rises faster at better sites, and somehow the bees at each site monitor their numbers there so that they know whether they've reached the threshold number (quorum) and can proceed to initiating the swarm's move to this site. This hypothesis can explain the cases of takeoff with disagreement among dancers as instances where a quorum was reached at one site before the competition between dancers from different sites had eliminated the dancing for all but one site.

We tested these two hypotheses with experiments performed on Appledore Island. In our first experiment, we presented four swarms, one at a time, with two identical nest boxes that offered the bees two superb nest sites. Our goal was to foster strong debates on our swarms and then see if they would take off before their dancing bees had reached a consensus (as Lindauer had reported for two of his swarms). In each trial, we positioned the swarm at the island's center, on a porch of the old Coast Guard building, and we placed both nest boxes near the rocky shore, each 250 meters (820 feet) from the swarm but in different directions, to the northeast and to the southeast. We also wanted to census the scout bees inside and outside each nest box, so each box was mounted against a window on the side of a lightproof hut (see fig. 3.11). The plan worked! We found that our swarms would discover both nest boxes at about the same time, would tend to develop a balanced debate over these two highly attractive sites, and would routinely take off when scout bees were still dancing strongly for both sites. Most telling was the spectacle that we witnessed on July 7, 2002. At 12:04 p.m., when both nest boxes were being advertised by dozens of bees performing vigorous dances, our swarm took off and then the large cloud of swarm bees split itself in two! Separate groups of airborne bees gathered on the north and south sides of the Coast Guard building, and at 12:09 each group began to move off slowly in the direction of "its" nest box. Both groups, however, traveled only about 40 meters (130 feet) toward their nest box before stopping, and at 12:15, when we noticed that the swarm's queen was back on the porch of the Coast Guard building, both groups started to return there and resettle around her.

This first experiment showed us that consensus among dancers is not necessary for a swarm to initiate its move to a new home, hence we could reject the consensus-sensing hypothesis for how scouts know when to start piping. This experiment also provided some support for the quorum-sensing hypothesis, because we noticed that swarms consistently started preparing for flight—that is, their scouts started piping—once 20 to 30 or more bees were seen together at one of the nest boxes, usually with about 10 to 15 bees inside and 10 to 15 outside. This suggested that in the decision-making system of honeybee swarms, 20 to 30 bees present simultaneously at one of the potential nest sites is a quorum. It should be noted, however, that because scout bees spend much of their time at the swarm cluster, seeing some 25 bees at a prospective nest site at any one time means that the total number of bees making visits to and thereby expressing support for this site is about 50 to 100 bees.

In performing our second experiment on Appledore Island, in June and July 2003, we sought to make a direct test of the quorum-sensing hypothesis for how scouts know when to start producing their piping signals. Our plan was to test a critical prediction of this hypothesis: delaying the formation of a quorum of scout bees at a swarm's chosen nest site, while leaving the rest of the decision-making process undisturbed, should delay the start of piping and thus the takeoff of the swarm. This was a critical prediction of the hypothesis because if we found that this prediction was wrong, then we would have dealt a deathblow to the hypothesis of quorum sensing.

Kirk and I devised a simple but effective way to delay quorum formation: we placed five desirable nest boxes close together at one location on the island

Fig. 7.14 Cluster of five nest boxes in one site on the eastern shore of Appledore Island. The swarm is 250 meters (820 feet) away, at the center of the island (to the right). The assistant, Adrian Reich, makes counts of the scouts outside each nest box.

(fig. 7.14). This caused the scouts visiting the site to be dispersed among five identical nest cavities rather than concentrated at one. We then saw how long it took a swarm, once it had discovered the site of the nest boxes, to start piping and eventually take off to fly to the site. We also performed with each swarm another control trial with just one nest box. The two trials for each swarm were performed using two different sites on the island, so each trial began in the same way, with one scout bee discovering an attractive nest box in a new site. In all four swarms that we tested, the scouts concentrated their attention on the single box in the one-nest-box trials and a crowd of bees built up there rapidly, but they distributed themselves evenly among the multiple boxes in the five-nest-box trials so the crowds of bees at these boxes built up more slowly. And in all four swarms, there was indeed a marked delay to start of piping and to start of takeoff in the five-nest-box treatment relative to the one-nest-box treatment. The times from discovery of the nest box(es) to start of piping and to start of takeoff were 162 and 196 minutes on average in the one-nest-box trials, but 416 and 442 minutes on average in the five-nest-box trials. It should be noted that the amount of waggle dancing back at the swarm did not differ between the two treatments. Also, the level of dance consensus was the same for both treatments; the bees were always unanimous in dancing only for the site of the nest box(es). It seems clear, therefore, that our five-nest-box treatment did not disturb anything in the decision-making process at the swarm cluster, and yet it delayed the start of piping and the start of takeoff. Thus this experiment yielded strong support for the quorum-sensing hypothesis.

Based on our two experiments conducted on Appledore Island in 2002 and 2003, Kirk and I drew the conclusion that a quorum of scouts at one of the proposed sites, not a consensus among dancers at the swarm, is the key stimulus for scouts to start piping and thereby initiate preparations for swarm takeoff. But how do we reconcile this conclusion with the fact that by the time the swarm takes off it *must* have a consensus among its scouts in order to fly as a unit to a single chosen site? One possible answer is that the preparations for takeoff, which generally take an hour or more, provide sufficient time for the positive feedback process of recruitment to the best site to produce the necessary unanimous agreement among the scouts. There may, however, be more to the story. It may be, for example, that the piping signals—which Kirk and I learned in 2006 are produced only by scouts returning to the swarm from the chosen site—inform the scouts from the losing sites (all those without a quorum) that the contest is over and that they should stop advertising these sites. This would certainly help the scouts reach full agreement about their future home, but whether the scouts from losing sites really respond to piping signals in this way remains unknown.

Exactly how the scouts sense a quorum also remains unknown. One possibility is that they use visual information. For humans, and perhaps also for bees, the constantly moving scout bees are easily detected visually outside the cavity and even inside it, at least around the entrance opening, which admits considerable light. Another possible means of sensing the number of bees at a site is by touch. It is a curious fact that as soon as a site acquires multiple scouts, they begin to make frequent contacts with one another. Many even start to perform what look like buzz-runs inside and outside the prospective nest site and so butt against other bees. It seems entirely possible that a bee could use the rate of contacts with scouts in general, or collisions with buzz-runners in particular, as an indicator of the number of fellow scouts at a site. Still a third possibility is the use of olfactory information. Scout bees standing in the entrance opening of a potential home often fan their wings and expose their scent organs—thereby releasing the lemon-scented blend of attraction pheromones, the message of which is "Come here!"—presumably to help other scouts find this special spot. It is possible that the level of these attraction pheromones rises with increasing numbers of bees at a site. Testing these various possibilities remains a subject for future study.

Why Quorum Sensing?

At first thought, it seems odd that the scout bees use quorum sensing rather than consensus sensing to know when to begin preparing their swarm for its flight to the new home. After all, a consensus among the dancers is needed for a swarm to execute successfully a move to its chosen homesite. Both Lindauer, with his Balcony and Moosach swarms, and Kirk and I with our Appledore swarms in 2002, saw what happens when a swarm takes off when its dancers are strongly divided between two sites: the cloud of airborne bees splits up, both halves stall in their moves, and finally they rejoin by resettling wherever their weary queen alights. Thus the bees have a big to-do but get nothing done.

Why don't the scouts use consensus sensing and thereby avoid the risk of their swarm splitting after takeoff and going nowhere? One likely reason is that sensing a consensus among the dancing bees would be extremely difficult for the bees. Presumably, each scout would have to poll the advertisements of her fellow scouts, which would involve traveling over the swarm cluster, reading dances, and keeping a mental tally of these readings. Doing all these things would be especially difficult on larger swarms with more scouts and thus more dances to poll. Quorum sensing, however, need not become more difficult with increasing swarm size, because the quorum size could be fixed, hence independent of swarm size.

Another likely reason that scouts don't use consensus sensing is that quorum sensing, unlike consensus sensing, enables a swarm to strike a good balance between speed and accuracy in its decision making. Consider first the matter of speed. Using a quorum as the trigger for the start of takeoff preparations means that these preparations can begin as soon as enough scouts have approved of one of the sites, even if many other scouts are still visiting and advertising other sites. In other words, there's probably no need to wait for full agreement if the outcome can be sensed in advance. If the bees used a consensus as the trigger, then the start of a swarm's takeoff preparations would be delayed by the extra time needed to reach a consensus. Consequently, the swarm bees would burn through more of the small store of energy (honey) they brought with them. This further depletion of a swarm's energy reserve would be considerable if the delay in start of flight preparations were to force a swarm to postpone its departure to the following day—swarms rarely take off after 5:00 p.m.—and so spend another chilly night camping out.

Now consider the matter of accuracy. It seems that the quorum used by the bees is 20 to 30 bees present simultaneously at a site (half inside, half outside), which requires that some 75 scout bees are actively supporting this site since each one spends only part of her time at the site. Using a 20- to 30-bee quorum evidently helps ensure accurate decision making because it guarantees that scout bees will not begin producing piping signals until a sizable number of them have independently scrutinized a site and judged it worthy of their support. This makes it extremely unlikely that a swarm will choose a poor site when a better one is available, for a poor site will not attract a (large) quorum of scouts. To see why, imagine that a scout makes a mistake—judging a poor site to be a good one—and recruits strongly to a poor site. Her followers will correct her error when they examine it themselves, find that it fails their scrutiny, and refrain from advertising the site further. Thus the mistake of the erring scout is soon silenced, the number of scouts at the site quickly dwindles, and the swarm rejects this low-quality site. I suspect that quorum size is a parameter of the bees' decision-making process that has been tuned over evolutionary time to provide an optimal balance between speed (favored by a small quorum) and accuracy (favored by a large quorum). We will examine this matter further in chapter 9.

The idea that a group can strike a better balance between speed and accuracy in its decision-making if it uses a quorum rather than a consensus to know when to start taking action is nicely illustrated by a story told to me by a neighbor friend who is a Quaker. Some years ago, the members of her Meeting wrestled with the question of whether or not to change the location of their meetinghouse. In meeting after meeting the subject was discussed, with the Friends always seeking a united wisdom, but every discussion ended with an adjournment for further consideration because an agreement could not be reached. Why? Because there was one Friend, an elderly lady, who felt strongly that the proposal was a mistake, withheld her consent, and so blocked a decision. If the Meeting had used a quorum, acting when a sufficient number or proportion of its members agreed, it would have reached its decision in a few weeks, but waiting for a consensus required four years. What finally enabled the Meeting to achieve a united judgment was the death of the one disapproving Friend. Some decisions do need to be made quickly, even if the choice is imperfect. The Quaker way of business, with its unceasing patience in finding full agreement, would be extremely risky for a homeless swarm of bees hanging in limbo under the open sky.

8
STEERING THE FLYING SWARM

Much have I marvelled at the faultless skill
With which thou trackest out thy dwelling-cave,
Winging thy way with seeming careless will
From mount to plain, o'er lake and winding wave.
—*Thomas Smibert,* The Wild Earth-Bee, *1851*

Thomas Smibert, writing about bumblebees flying home "o'er lake and winding wave" in his native Scotland, lauded the marvelous ability of bees to return home after visiting distant flowers. His praise is richly deserved. We now know that a worker honeybee can navigate to and from flowers blooming 10 or more kilometers (more than 6 miles) from the hive, a thoroughly respectable distance for a creature only 14 millimeters (about half an inch) long. We also now know that bumblebees and honeybees find their way home using navigation methods like those used for ages by sailors making a passage over open water to reach a familiar harbor: steering according to a compass—for bees, this is the sun—and keeping a running tally of distance traveled, but then relying on memorized landmarks when within sight of the goal. How individual bees can range so widely without getting lost was one of the mysteries that Karl von Frisch and Martin Lindauer dug into most deeply in the 1950s, and since then other biologists have further revealed how a bee guides herself out to flowers and home to hive.

Meanwhile, the related mystery of how a swarm of bees steers itself to its new home was neglected, probably because it seemed a mind-boggling puzzle. Somehow, a school-bus-sized cloud of some ten thousand flying insects manages to sweep straight from bivouac site to new dwelling place. The path of its flight usually stretches for hundreds or thousands of meters (up to several miles), traversing fields and forests, hilltops and valleys, and swamps and lakes. Perhaps most amazingly, the airborne colony pilots itself over the countryside to one specific point in the landscape: a single knothole in one particular tree in a certain patch of forest. And as the group closes on its destination, it gradually lowers its flight speed so that it stops precisely, and gracefully, at the front door of its new home. How do ten thousand bees accomplish this magnificent feat of oriented group flight? In the past few years, with the introduction of digital video technology, it has become possible to perform the sophisticated data collection and image processing needed to track individual bees in a flying swarm and thus unravel the mechanisms of flight guidance in honeybee swarms. In this chapter, we will look at these mechanisms and we will see that the scout bees, yet again, play the leading role in our story.

Swarm Chasers

In the summer of 1979, I returned to my family home in Ithaca, New York, to work again with my first mentor and good friend, Roger "Doc" Morse, professor of apiculture at Cornell. A few years before, Doc and one of his students, Alphonse Avitabile, now professor emeritus at the University of Connecticut, had found that when a swarm of bees flies to its new home, the workers continuously monitor the presence of the queen within the swarm cloud by smelling the "queen substance" pheromone that wafts steadily from her body. This pheromone, the major component of the material secreted by the mandibular glands in the queen's head, is a 10-carbon fatty acid whose exact name is (E)-9-oxo-2-decenoic acid (I will call it simply 9-ODA). If the bees in an airborne swarm keep smelling this particular chemical substance, they will keep flying toward their new home address, but if they don't catch its aroma, because their queen has dropped out to rest, they will cease flying forward, mill about until they find their missing queen, and then cluster around her wherever she has alighted. Sooner or later, the swarm will again take off and proceed to its destination. Clearly, the workers in a flying swarm take great care to avoid losing their all-important queen.

To test whether 9-ODA is the critical indicator of the queen's presence, Doc and Al Avitabile had conducted a slightly evil experiment. They had set up artificial swarms so that each swarm's queen was imprisoned in a small cage and then, when each swarm had finished its house hunting and was taking off to move to its chosen site, they had painted 9-ODA on the backs of five worker bees as they launched into flight. Each swarm treated in this way took off, flew out of sight, and . . . never returned! Swarms that were treated identically except that no workers were painted with 9-ODA also completed the takeoff process, but they flew off only about 50 meters (150 feet) before they returned and resettled around the caged queen. Clearly, the presence of a bee bearing the special fragrance of 9-ODA suffices to convince an airborne swarm that its queen is on board. To this day, I feel sadness for the orphaned swarms produced in this otherwise superb experiment.

In watching his 9-ODA-treated swarms fly out of sight, Doc became intrigued by how they conduct their flights, and in 1979 he invited Kirk Visscher (just starting graduate studies with Doc) and me to help him tackle this problem. Our first goal was simply to watch a swarm perform a flight, from start to finish. To do so, we went to Appledore Island, where we knew we could control a swarm's flight path. We took with us a medium-size (11,000-bee) swarm, and we carefully positioned it and a nest box on the island so that we would be able to run beneath the swarm throughout its flight. The jungles of poison ivy on Appledore limited us to running along its roads and trails, none of which is straight, but we managed to find a sufficiently linear "track" 350 meters (1,150 feet) long that would enable us to stay close to our swarm throughout its journey. At one end we placed the swarm, at the other end the nest box, and every 30 meters (100 feet) in between a flagged stake. By noting when the center of the flying swarm passed over each stake, we could later calculate its speed during each stage of its flight.

As expected, a scout from our swarm soon found our nest box and the bees dancing for it quickly dominated the scout bees' debate. While we were waiting for the scouts to finish their deliberations, we applied a dot of blue paint to every bee that performed a dance for the nest box, and we noted every five minutes the percentage of bees visible at the nest box that were marked with blue. Knowing that we painted 143 scout bees and finding that on average 29 percent of the scouts at the box were painted, we estimated that approximately 495 bees ($143 = 0.29 \times 495$) had visited the nest box before takeoff. Thus we learned that fewer than 5 percent of the 11,000 bees in our swarm were familiar with its destination upon takeoff.

Equally interesting is what we learned about swarm flight speeds once the swarm became airborne. We saw that the cloud of swarming bees hung over the bivouac site for about 30 seconds, then began moving slowly off in the direction of the nest box. It covered the first 30 meters (100 feet) at less than 1 kilometer per hour (about 0.5 miles/hr) but accelerated steadily so that after 150 meters (500 feet) it had reached its top speed of 8 kilometers per hour (5 miles/hr). What was most surprising was the way that the swarm somehow managed to

Fig. 8.1 Kirk Visscher (left) and the author (right) in 2006, watching a swarm in the process of moving into a nest box on Appledore Island.

apply its brakes before it reached the nest box. Starting about 90 meters from the box, it gradually trimmed its speed and finally came to a halt with the center of the swarm cloud less than 5 meters (15 feet) from its goal. Over the next two minutes, the scout bees appeared in increasing numbers at the box's entrance opening—5 after 20 seconds, 40 after 50 seconds, and over 100 after 90 seconds—releasing Nasonov gland pheromones to show the ignorant bees the way into their new residence. Within three minutes of the swarm stopping before the nest box, the bees were landing so heavily on it that they blanketed its front. Soon they were marching inside en masse, creating a whirlpool of bees that wheeled slowly around the entrance hole (fig. 8.1). The queen slipped in without fanfare six minutes later, and before 10 minutes had elapsed since the swarm's arrival, nearly all the bees were safely inside their new home.

I took a liking to chasing swarms that day, but didn't follow up on our observations until 25 years later, in the summer of 2004, when I had the immense good fortune of being joined by Madeleine Beekman, a behavioral biologist from the Netherlands. Madeleine had recently completed postdoctoral studies in England with my friend Francis Ratnieks, a noted bee expert, and had become intrigued by the mystery of swarm flight guidance. She joined me for a summer of swarm studies at Cornell and turned out to be the best possible collaborator: intelligent, hard working, and good-natured. She is now on the faculty of the University of Sydney in Australia.

The observational setup on Appledore Island had been rather crude and we looked for ways to improve it. We wanted to describe the flight behavior of swarms more precisely and to perform controlled experiments. We decided the way to do this was to fly swarms across the large meadow beside my laboratory at the Liddell Field Station, just off the Cornell campus. In the middle of this 26 hectare (65 acre) expanse of grass stands one large ash tree (*Fraxinus americana*) with spreading limbs, which provided a perfect place for hanging the nest box that we wanted our swarms to choose for their new home. Of course, there were attractive natural nesting cavities in the woods surrounding this field and beyond, but I had learned from my studies on Appledore that if I watched a swarm with infinite patience and plucked from it every scout bee performing a dance for some site other than my nest box, I could keep the swarm's attention fixated

Fig. 8.2 Madeleine Beekman beneath a flying swarm moving to the right. The orange marker she is holding aloft is a size reference 45 cm (18 inches) long. Note the longer streaks in the top of the swarm cloud, indicating bees flying at high speed.

on the nest site I was offering. This worked well. We flew many swarms along a 270-meter (886-foot) flight path that ran from near the laboratory out to the ash tree. We divided this flyway into 30-meter (98-foot) segments so we could make measurements of swarm flight speed. Also, to make accurate measurements of the dimensions of swarm clouds upon takeoff, we created a 20 x 20 meter (66 x 66 foot) "launch pad" for our swarms. This was a closely mowed area that was gridded with stakes spaced 4 meters apart and in which we erected a 6-meter-tall pole with 1-meter markings. Each swarm was set up in the launch pad's center, and upon takeoff we measured its cloud's length and width with reference to the grid, and the heights of its cloud's top and bottom with reference to the pole. We also photographed the flying swarms from the side for later analysis of the movement patterns of individual bees.

We began by watching the flights of three swarms, each containing approxi-

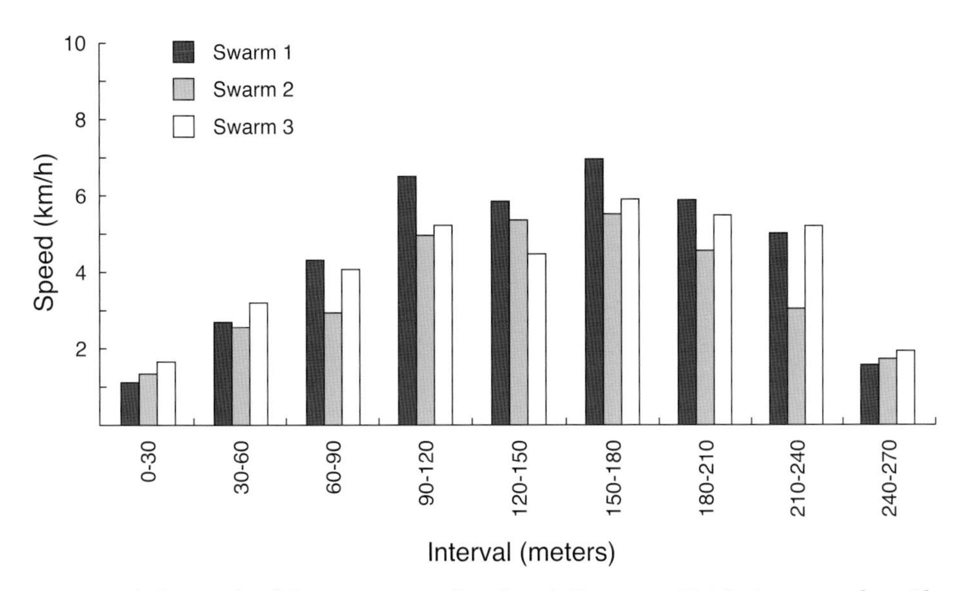

Fig. 8.3 Flight speeds of three swarms as they flew 270 meters (886 feet) to a nest box. Their top speeds were 5 to 7 kilometers per hour (3 to 4 miles per hour). When swarms perform longer flights they can reach speeds of nearly 12 kilometers per hour (over 7 miles per hour).

mately 11,500 bees, which is the median size of natural swarms. In launching itself into flight, each clustered swarm exploded into a buzzing cloud of bees some 10 meters (33 feet) long, 8 meters (26 feet) wide, and 3 meters (10 feet) tall. The bottoms of these swarm clouds swirled about 2 meters (6 feet) from the top of the meadow grass, hence (thankfully) a bit over our heads! Knowing these dimensions, we calculated that within each swarm cloud the flying bees were spaced, on average, only about 27 centimeters (10 inches) apart, which means that they were functioning at a density of some 50 bees per cubic meter (1.4 bees per cubic foot) (fig. 8.2). Amazingly, the bees rarely if ever suffered midair collisions.

The flight patterns of all three swarms matched what Doc, Kirk, and I had seen with our swarm on Appledore Island. Each swarm at first moved very slowly, then smoothly accelerated to a top speed of about 6 kilometers per hour (4 miles per hour), and finally braked gently, coming to a full stop at its new home (fig. 8.3). And as before, we noticed a brief delay between when each swarm had reached its destination and the moment its scouts began settling at the entrance hole and releasing Nasonov gland pheromones, but that once this chemical signal was being discharged, the rest of the swarm cloud quickly settled on the nest box. The bees didn't hesitate to move inside and within 10 minutes nearly all had disappeared indoors. Each swarm executed the entire migration process—takeoff, flight, landing, and entry—with precision and so completed it in less than 15 minutes.

Leaders and Followers

What makes the precisely oriented flight of a honeybee swarm to its new home so wondrous is that only a small percentage of its members know the swarm's travel route and final destination. As mentioned already, less than 5 percent of the swarm that Doc, Kirk, and I studied on Appledore Island had visited the nest box, and so knew its exact location, before the swarm made its flight to its new home. This finding was later confirmed by Susannah Buhrman and me in the study in which we prepared swarms of individually labeled bees, video recorded the scouts' dances, and determined which proposed nest site each scout advertised. In all three swarms that we studied, only 1.5 to 1.7 percent of the bees performed dances for the chosen site. This figure, combined with the figure of approximately 50 percent for the fraction of the scout bees from a high-quality site that advertise it with a dance (what Kirk and I found in our study of how scouts encode nest-site quality in their dances; see chapter 6), yields the estimate that only 3 to 4 percent of the bees in each swarm had visited the chosen site and so knew the exact location of their swarm's future dwelling place. Clearly, when a swarm flies to its new home, it relies on a relatively small number of informed scout bees—approximately 400 individuals in an average-size swarm of 10,000 bees—who function as guides or leaders of all the rest. How does this system of leaders and followers work?

Three hypotheses have been proposed for how an informed minority guides the ignorant majority when a swarm flies to its new home. The first hypothesis suggests that the transfer of information from leaders to followers relies on a chemical signal. In their 1975 paper on how the workers in a swarm perceive the presence of their queen by sensing the 9-ODA she releases, Al Avitabile, Roger Morse, and Rolf Boch proposed that scouts provide flight guidance by means of

Fig. 8.4 Worker bee exposing her scent organ, to release attraction pheromones, by tipping down the last segment of her abdomen.

the attraction pheromones produced in the Nasonov gland that is part of the scent organ at the tip of a worker bee's abdomen (fig. 8.4). Their idea was that scouts might discharge these pheromones along the front of the swarm cloud to attract, and thereby guide, the nonscouts to move in this direction.

The other two hypotheses suggest that the information transfer from informed bees to ignorant bees works by vision instead of olfaction. One hypothesis, called the "subtle guide hypothesis," was proposed in 2005 by a team of biologists from Princeton University in the United States and the universities of Leeds and Bristol in the United Kingdom: Iain Couzin, Jens Krause, Nigel Franks, and Simon Levin. According to this hypothesis, the informed bees do not conspicuously signal the correct travel direction; instead, they steer the swarm simply by tending to fly in the direction of the new home. By making computer simulations of airborne swarms, the authors showed that if each bee in a flying swarm (1) attempts to avoid collisions by turning away from any neighbors within a critical distance, (2) tends to be attracted toward and aligned with neighbors outside the critical distance, and (3) flies either with a preferred movement direction (informed bees) or without a preferred movement direction (ignorant bees), then the swarm will be steered toward its new home even if the proportion of informed individuals is very small. Remarkably, in large groups, like honeybee swarms, this proportion can be less than 5 percent. It is an intriguing hypothesis, for it shows that the bees in a flying swarm might not need to know which of them know the travel route, hence are the leaders.

The second vision-based hypothesis, called the "streaker bee" hypothesis, was sketched out in 1955 by Martin Lindauer. At the very end of his magnum opus on house hunting by honeybees, Lindauer reported having seen "that several hundred bees, in more rapid flight, always shoot forward toward the front of the swarm cloud, that is to say in the direction of the nest site. While the swarm cloud slowly continues its flight in this direction, these guiding bees slowly fly back along the border of the swarm cloud and again shoot to the front in rapid flight." The streaker bee hypothesis suggests that the informed bees conspicuously signal the correct travel direction by repeatedly making high-speed flights through the swarm cloud. (Note: according to the subtle guide hypothesis, the informed and ignorant bees fly at the same speed.) In the streaker bee hypothesis, the ignorant bees behave as is suggested for the subtle guide hypothesis except for one thing; rather than align themselves with neighbors in general, the ignorant bees preferentially align themselves with fast-flying neighbors. So the two key differences between the subtle guide and streaker bee hypotheses are whether or not the informed bees (leaders) point the way with high-speed flights and whether or not the ignorant bees (followers) favor alignment with fast-flying bees. Computer simulations, similar to those done for the subtle guide hypothesis, have shown that the streaker bee hypothesis is a plausible mechanism of swarm flight guidance. Thus both the subtle guide and streaker bee hypotheses for swarm flight guidance are a possibility. The burning question is which, if either, is the reality.

Scent Organs Sealed Shut

After describing the flights of swarms across the meadow at the Liddell Field Station, Madeleine Beekman and I set ourselves the goal of testing the hypothesis that scout bees guide their swarm using the attraction pheromones produced in their scent organs. We knew we had to prepare swarms in which every worker had her scent organ sealed shut and then see if these swarms could perform a well-oriented, full-speed flight down our flyway at Liddell.

The scent organ of a worker honeybee lies on the dorsal surface of the abdomen, at the front edge of the last abdominal segment. It consists of several hundred gland cells (the Nasonov gland, named after the Russian scientist who first described it, in 1883) whose ducts open onto the membrane that connects the last two plates ("tergites") covering the top of the abdomen (fig. 8.5). The secretion—which consists mainly of citral, geraniol, and nerolic acid and has a pleasant lemony aroma—collects on this membrane. Usually this area is concealed by the overlap of the two tergites, but if a worker bee bends the apical segment of her abdomen downward, she exposes the membrane and releases the scent. Using a fine paintbrush, one can paint over the joint between the last two tergites, and when the paint dries the two tergites will be stuck together so that the treated bee can no longer expose her scent organ.

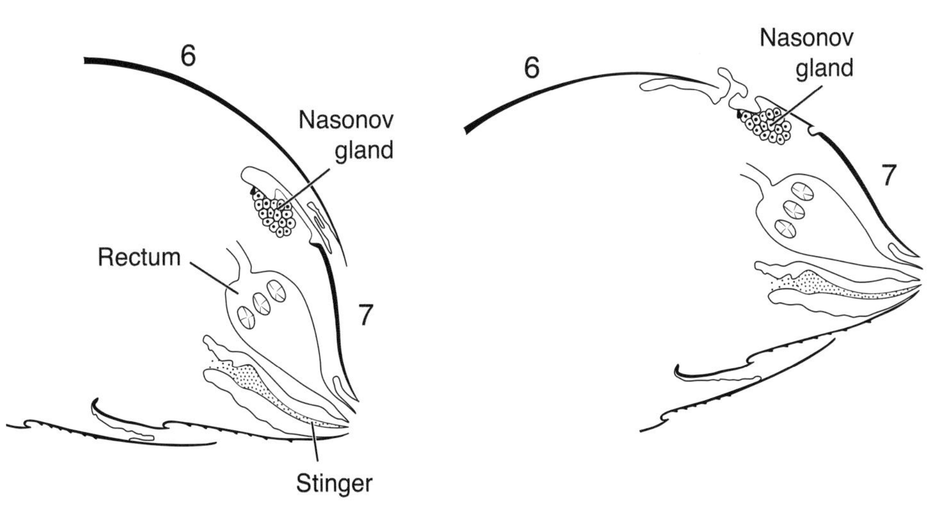

Fig. 8.5 Sections through the abdomen of a worker bee, showing (left) the scent organ closed in the rest position and (right) the scent organ exposed by raising the abdomen and tipping the last abdominal segment downward. 6 and 7, sclerotized integument (tergites) atop the 6th and 7th abdominal segments.

We tried various paints. The first ones tended to crack easily so that after a few days many bees were able to release the balmy Nasonov gland pheromones from the scent organ, but eventually we found that Testors gloss enamel paint makes a lasting seal of a bee's abdominal segments. At this point we adopted a standard procedure for preparing our test swarms. After immobilizing groups of 10 to 20 bees in plastic bags in a refrigerator, we put one group of these chilled bees at a time on ice, painted over each bee's scent organ, and then poured the still immobile bees into a screened cage with their queen where they would warm up and become part of the artificial swarm we were creating. We did this over and over until we had 4,000 bees properly painted, enough for a small "treatment" swarm. To control for any effects of chilling, painting, and handling, we also prepared 4,000-bee "control" swarms in which we did everything the same except that we applied the dot of paint to each bee's thorax instead of her abdomen.

We eventually flew six swarms, three treatment swarms, and three control swarms. The two types, when airborne, formed similar-sized clouds of flying bees (8 meters long, 8 meters wide, and 3 meters high). Most importantly, both treatment and control swarms flew *directly* and *quickly* to the nest box! As we had seen previously with the large swarms, these small swarms accelerated steadily for the first 90 meters, reached peak flight speeds after flying 90 to 120 meters, started slowing down after flying 210 to 240 meters, moved very slowly during the final 30 meters, and finally stopped at the nest box. The maximum speeds of the treatment swarms were 6.8, 3.6, and 6.8 kilometers per hour, while those of the control swarms were 6.7, 6.4, and 7.2 kilometers per hour. (The second treatment swarm flew more slowly than the others because it flew against a fierce headwind, whereas all the rest encountered at most a slight breeze.) There was, however, one important way in which the two types of swarms did behave differently: once they reached the nest box, the treatment swarms took much longer than the control swarms (20 minutes versus 9 minutes, on average) to move into the box. Why? Almost certainly, it was because the scouts in the treatment swarms were unable to help the nonscouts find the entrance to their new home by marking it with Nasonov gland pheromones. They certainly tried hard to do so. The scouts landed at the three entrance holes in the nest box and stood there flamboyantly with abdomens elevated and wings whirring, but they could

Fig. 8.6 Bees with their scent organs sealed shut standing on the front of a nest box and attempting to release Nasonov pheromones.

not bend down the last abdominal segment to expose the scent organ (fig. 8.6). (To be sure about this, we inspected 250 bees from each swarm shortly after it entered the nest box and found that only a miniscule percentage of the bees, less than 1 percent, had cracked paint seals.) Because our treatment swarms executed their flight plans as flawlessly as our control swarms, except during the landing phase when the scent organ clearly plays an important role, we concluded that the informed scouts don't provide their ignorant sisters with flight guidance information using the Nasonov gland pheromones.

Streams of Streakers

Madeleine and I next started to test the streaker bee hypothesis. We believed that we had seen, while watching our swarms sweep across the meadow to the nest box, what Lindauer had reported seeing in an airborne swarm. Most bees fly about within the cloud in rather slow and looping flights, but a few shoot straight through the cloud in the direction of the swarm's new home. Also, it looked to us like the streaking bees zoomed mainly through the top of the swarm cloud. But we were not 100 percent confident of our sightings and certainly we had no hard data, so we decided to try to use conventional still photography to get solid information and check our impressions. Using a 35 mm camera, color transparency film with a slow film speed (DIN 64), and a moderately long exposure time (one thirtieth of a second), we found that if we photographed a flying swarm from the side and under a clear sky, we could get photos that "captured" the entire swarm cloud and in which individual bees appeared as small, dark streaks on a bright background (see fig. 8.2). The length of each bee's streak indicated her flight speed, and the tilt of her streak indicated her flight angle relative to horizontal, the orientation of level flight. These photos showed unambiguously that a small minority of the bees in an airborne swarm do whiz through it at the maximum flight speed of a worker bee—about 34 kilometers per hour (20 miles per hour)—and that all the rest of the bees buzz along much more slowly. We also found that the streaks of the fast-flying bees, compared to those of the slow-flying ones, are more apt to be horizontal, indicative of straight and level flights. Finally, we gleaned from the photos the finding that the speeding bees, the streakers, do indeed operate mainly in the top of the swarm cloud. This makes sense, assuming that these bees are providing flight direction information to the other bees, for by streaming over their sisters the fast-flying bees position themselves where they are easily seen against the bright background of the sky.

Computer Vision Algorithms for Tracking Bees

The photographic study that Madeleine and I made in 2004 gave support to the streaker bee hypothesis, but it was not a rigorous test between this hypothesis and the subtle guide hypothesis. This is because our photographs, taken from the side of a flying swarm, could not tell us the flight directions of the fast-flying bees (whether toward the new home site, away from it, or some angle in between). And the key to resolving the subtle guide and streaker bee hypotheses is knowing whether or not the flights of the speedsters in a swarm point mainly toward the swarm's new home. The two hypotheses make distinct predictions about this matter. The subtle guide hypothesis predicts that the fast-flying bees *will not be* heading mainly in the direction of the new homesite because, according to this hypothesis, the informed bees don't signal the travel direction with high-speed flights through the swarm cloud. In contrast, the streaker bee hypothesis predicts that the fast-flying bees *will be* heading mainly in the direction of the new homesite because, according to this second hypothesis, this is how the informed bees share their knowledge of the swarm's travel direction. Some of the speedy bees will be informed bees indicating which way to go and probably some will be ignorant bees reacting to the informed bees.

In 2006, when it became possible to track individual bees in a flying swarm and measure each bee's position, flight direction, and flight speed, it became clear that the high-speed fliers in a swarm are indeed zipping toward the swarm's new home. So now it seems clear that the streaker bee hypothesis is correct. The two people who were instrumental in developing the tools for tracking individual bees in an airborne swarm are Kevin Passino, professor of electrical and computer engineering at Ohio State University, and his brilliant graduate student, Kevin Schultz.

One of the great benefits of the academic life is that it gives you opportunities to visit other universities and meet remarkable people, some of whom share your intellectual excitement about a particular mystery. I met Kevin Passino on a trip to Ohio State University in the spring of 2002. I was there on a lecture engagement, not to enlist an engineer for collaborative work, but upon meeting Kevin I sensed he was a marvelous engineer with precisely the right inclination for a joint scientific venture. Here was someone who devised automatic control systems for technological applications, but who also liked to look at biological systems for inspiration. I was to learn later that "biomimicry" is a hot approach among control engineers, for the methods of automatic control in living organisms are exceptionally powerful and robust, having been tested and tuned by natural selection for millions of years. As I recall, the outcome of our first meeting was an agreement that we would team up. Already we had found splendid common ground on the puzzles of forager force allocation and swarm flight guidance. In the words of Kevin, the honeybees have evolved a "cooperative control strategy for groups of autonomous vehicles," and he was keen to join in exploring it.

Once Madeleine Beekman and I had disproved the pheromone hypothesis for swarm flight guidance and had published the results of our simple photographic analysis that confirmed the existence of streaker bees in swarms, Kevin realized that what was needed next was to video record an airborne swarm from below using a high-definition video camera. He had a hunch that by tapping into the latest video technology, especially point-tracking algorithms invented by engineers working on computer vision, it should be possible to track individual bees as they flew over the video camera and determine for each bee her position in the swarm cloud, her flight speed, and her flight direction. So Kevin purchased the necessary camera and joined Kirk Visscher and me during our field trip to Appledore Island in the summer of 2006. We set up a swarm by the old Coast Guard building at the island's center, placed an attractive nest box on the eastern shore 250 meters (820 feet) away, and charged up the camera's battery. Our goal was to video record the swarm as it flew over the camera at two points along its flight path: 15 meters (50 feet) from the bivouac site, when the swarm had just taken off and was still moving slowly, and 60 meters (200 feet), when the swarm was well under way and had picked up considerable speed. The camera was equipped with a wide-angle lens, so its field of view included most of the width of the swarm cloud, though not its entire length. The camera also had an extremely high shutter speed—one ten thousandth of a second—so in each frame of the video recording a bee appeared as a short blob, not a long streak. Our biggest obstacle to success that summer was the wind sweeping over Appledore Island, which made it difficult to get swarms to fly directly over our camera. In still air a swarm will course along a predictable beeline to its new home, but in windy air the track of a swarm's flight is wildly unpredictable as chaotic gusts push the airborne bees about, knocking them off a direct line of travel. And Appledore Island, anchored in the Atlantic Ocean six miles off the southern coast of Maine, is thoroughly windblown. It is so much so that in 2007 the Shoals Marine Laboratory erected on the island a 27.5-meter (90-foot) tall wind turbine to harvest some of the wind's energy. Now the laboratory acquires a substantial portion of its electrical power from this limitless source. On June 29 and July 2, 2006, however, we were blessed with two days of wonderfully calm air, and twice we got a swarm to fly directly over the video camera at both the 15-meter mark and the 60-meter mark along the line running straight to the bees' new home.

With these two sets of recordings of swarm flyovers "in the can," Kevin Passino had material for his PhD student, Kevin Schultz, to tackle. Over the next two years, Kevin S. created a computer algorithm that semiautomated the data-gathering process. In essence, the procedure involves examining each ellipsoidal blob (bee image) in a given video frame, noting its orientation (the angle between the major axis of its ellipse and the bottom edge of the video frame), and then pairing it up with the blob on the next video frame that represents the same bee. The pairing process involves finding for a given blob in the first frame the blob in the second frame that best matches its position and orientation. This process is repeated with the blobs of the second frame being paired with blobs of the third frame, and so on, to build up, frame by frame, detailed trajectories of individual swarm bees as they flew across the video camera's field of view back on Appledore Island. The size of a blob—the length of the major axis of a blob's ellipse—indicates the height of the bee above the camera. So the bees in the top and bottom portions of the swarm cloud were distinguished, and it was even possible to make three-dimensional reconstructions of the individual bees' flights. What a tour de force!

It is hard to convey in words what it is like to go from watching thousands and thousands of swarm bees swirling over head in seeming random motion, to seeing graphs that show wonderfully clear patterns in their movements. Every detail of the swarm bees' collective motion was a complete revelation, for before Kevin P. and Kevin S. devised their process of blob tracking on digital video recordings, no one could even begin to see these patterns. The human visual system is a stupendous biological computer, capable of amazing feats of information processing such as instantly recognizing a face not seen in years, but even it is overwhelmed by so many bees moving so rapidly and so wildly.

The most important pattern revealed by the analysis of the video recordings is that the fast-flying bees were indeed streaking in the direction of the chosen homesite. Looking at figure 8.7, which shows the individual bees' flight speeds in relation to flight direction, we can see that the speediest bees were zooming directly toward the new home, while the slowest ones were heading in the opposite direction. By comparing the plots for the top and bottom portions of the swarm cloud, we can also see that the speedsters were mainly in the top portion of the swarm, confirming what Madeleine and I had detected with our side-view photographs of moving swarms. A third important feature of the plots shown in figure 8.7 is that the peaks in flight speed are higher at the front section of the elongated swarm cloud than at its rear (true for both the top and the bottom

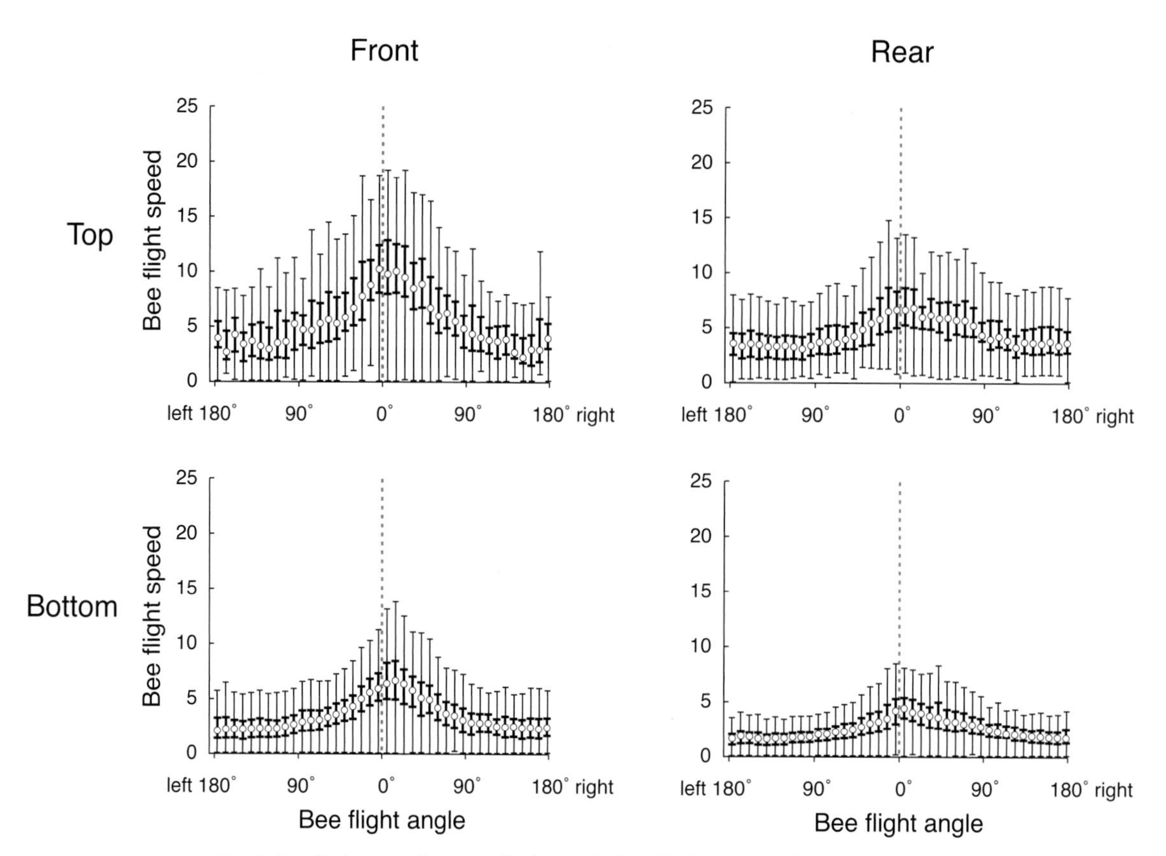

Fig. 8.7 Flight speed versus flight angle for the bees in a flying swarm when it had flown 15 meters (50 feet) from its bivouac site. Bees with a flight angle of 0° were flying straight toward the new home; all the rest were flying at some angle to the right or left of the homesite direction. The measurements of the flying bees are organized into four regions: front and rear of both the top and bottom layers of the swarm cloud. The bold portion of each vertical line denotes the range of the 50 percent of the measurements closest to the median value (indicated by the open circle in each line) and the thin portion denotes the range of the 50 percent of the measurements farthest from the median value. The units of flight speed are bee lengths per video frame.

of the swarm cloud). This shows that the fastest bees tended to be in the front of the swarm. A painstaking analysis by Kevin S. of the velocities of individual bees found that not only do bees that are flying in the direction of the nest site tend to fly with the highest velocities, but also they tend to accelerate (increase their velocities) as they move from swarm rear to swarm front. It seems likely that some of this rise in flight speed comes about as the ignorant/follower bees "latch on" to the informed/leader bees, boosting their speed as they chase after the streaker bees. If so, the information about flight direction (expressed in the flight direction of the fast-flying bees) and the boosting of flight speed are likely to spread from the informed bees to some of the ignorant bees who, through their own faster flights, will start to influence other ignorant bees. This chain reaction of informed/leader bees begetting more informed/leader bees could lead to a widespread induction of bees to fly toward the nest box and to fly faster. This may explain the increase in overall swarm speed over time that is shown in figure 8.3, and that is so impressive to any beekeeper who tries to follow a fugitive swarm to its new home by running along beneath it.

The discovery that the bees flying toward the new home are traveling far faster than the other swarm bees led Kevin Passino, Kevin Schultz, and me to conclude that streaker bees, not subtle guides, appear to provide the flight guidance to an airborne swarm of honeybees (fig. 8.8). We would like, however, to test the streaker bee hypothesis more rigorously by performing an experiment analogous to the sealed-scent-organ test of the attraction pheromones hypothesis: block the proposed means of guidance and see if this renders a swarm incapable of making a well-oriented, full-speed flight to its chosen destination. Unfortunately, nobody has succeeded yet in figuring out how to prevent the informed bees from performing high-speed flights. Madeleine Beekman tried trimming the wingtips of scout bees by a millimeter or so, a manipulation that is known to reduce a bee's maximum flight speed, but she found that this surgery also caused her bees to stop scouting. Maybe some other approach will do the trick. Glue small airfoils or short strings to scout bees to increase the drag they experience during flight? Find bees that have a genetic mutation that causes them to fly slowly? Anyone who figures out a way to prevent streaking by scout bees will have set the stage for a beautiful experiment.

In the meantime, Madeleine Beekman and two students, Tanya Latty and Michael Duncan, have succeeded with a different approach to testing the streaker bee hypothesis. They performed an ingenious experiment in which they caused numerous fast-flying forager bees to zoom through an airborne swarm in a direc-

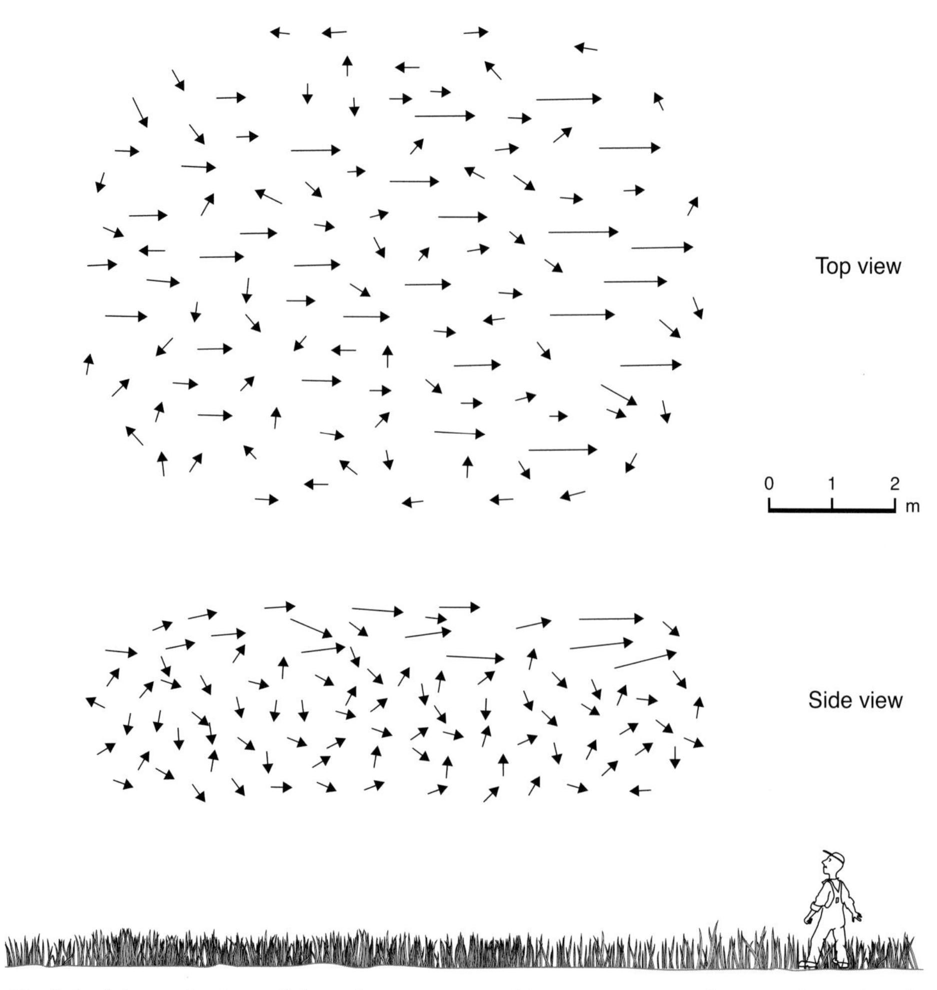

Fig. 8.8 Schematic view of the velocity vectors of bees in a swarm flying to the right. The streaker bees are mainly in the top of the swarm cloud. Velocity vectors are shown for only a small fraction of the bees in the swarm.

tion perpendicular to the swarm's intended flight path (fig. 8.9). If the streaker bee hypothesis is correct, the foragers sweeping in from the side should create conflicting directional information in the swarm and thereby disrupt its flight guidance. This is exactly what was found. Of six test swarms that attempted to fly to a nest box 100 meters (330 feet) away when foragers were zipping back and forth across the swarms' flight path, only one reached the nest box intact,

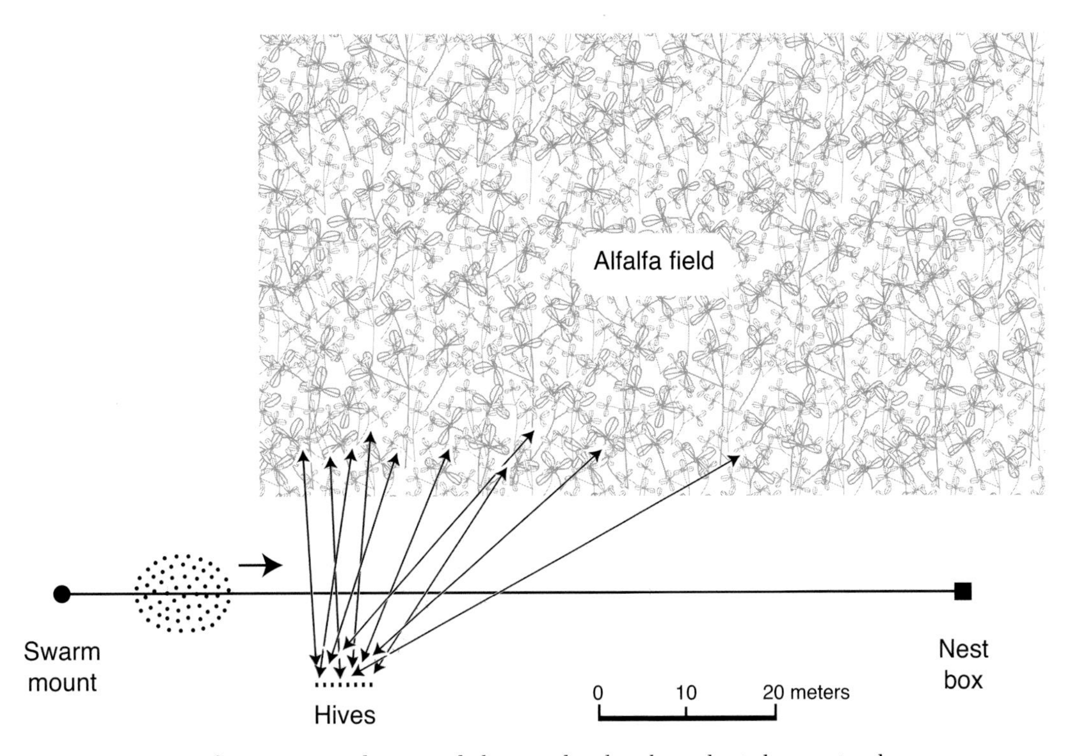

Fig. 8.9 Layout of experiment that tested the streaker bee hypothesis by causing bees to zoom through an airborne swarm at a right angle to the swarm's flight direction, and then seeing if this disrupted the swarm's flight guidance. The cross-flying bees were foragers that were racing out from a battery of eight hives to gather rich nectar from a nearby field of alfalfa (*Medicago sativa*).

and even this swarm was knocked temporarily off course. The other five swarms all started out, as usual, by moving straight toward the nest box, but upon hitting the "forager highway" they either fragmented or veered widely off course. In contrast, when the experimenters flew four control swarms—ones identical to the test swarms but without the cross traffic of foragers—they saw each one stay together and fly directly to the nest box. Clearly, the heavy traffic of foragers crossing the flight path of the test swarms disrupted their mechanism of flight guidance, evidently by injecting misleading visual information into the clouds of swirling bees.

Assembling the Flight Navigators

There are many questions left unanswered about the remarkable flights of honeybee swarms. How does the moving group apply the brakes when it arrives within about 100 meters (330 feet) of its new residence? Also, how do the informed bees make their repeated streak flights through the swarm cloud? Do they stop when they reach the front and let the other bees fly past, or do they fly inconspicuously rearward along the swarm's bottom, where they may be nearly invisible against the underlying vegetation? There is also the mystery of how a swarm makes sure that it is fully stocked with bees who know its flight route before it begins its journey to the new home.

It is striking how virtually all the scout bees who have visited the chosen homesite, and so can steer the airborne swarm to it, abandon the future dwelling place and assemble on the swarm cluster shortly before it launches into flight. I first witnessed this phenomenon in August 1974, shortly before I started graduate school, when I first watched a swarm go through its house-hunting process. I had set up an artificial swarm and a nest box behind my parents' house, in an abandoned field dominated by blooming goldenrod plants (*Solidago* spp.) and young white pine trees (*Pinus strobus*), one of which supported my box for the bees. To my great good fortune, the scout bees chose my humble plywood box for their future home. Soon I was dashing back and forth along the 150-meter-long (500-foot-long) path between swarm cluster and nest box, doing my best to watch both the growing party of excited dancers on the swarm and the strengthening throng of scout bees scrutinizing the nest box. Part way through the afternoon, I was shocked to see a sudden drop in bees at my box. On my previous visit, 15 minutes earlier, I had counted some 25 bees examining the box, but now I saw only two or three bees, and in a few more minutes the place was totally deserted. This collapse in the scout bees' interest baffled me, until I glanced back toward the swarm's bivouac site and saw my swarm, now a diffuse ball of swirling and shining bees, "rolling" straight toward me over the sunny field. Evidently, the scouts had abandoned the nest box to be in the swarm cluster at its moment of departure.

Since then, when performing experiments on swarms, I have come to rely on the conspicuous drop in scouts at the nest box as a reliable indicator that the swarm has completed its decision making and is about to take off (see figs. 5.5 and 5.7). It certainly makes sense for the scouts to assemble at the swarm shortly before departure, for we have seen how only 3 to 4 percent of a swarm's bees know its flight plan, and with such a small minority of navigators it is probably important to have as many as possible on board. But precisely how this is achieved remains a mystery. Does the gathering of the scout bees on the swarm arise simply by these bees returning to the swarm as usual and then lingering there when they detect one of the flight initiation signals, either worker pipings or buzz-runs? Or might their assembly on the swarm be triggered by them hearing, feeling, seeing, or smelling an unknown signal of impending departure that is produced at the nest box? I wouldn't be surprised if the bees possess some secret gadgetry for ensuring that a swarm about to take flight is well stocked with the informed bees who can pilot it safely to its new home.

9

SWARM AS COGNITIVE ENTITY

I'm a systems neurobiologist who studies how the three pounds of goo we call a human brain makes decisions.
—*William Newsome, 2008.*

The previous six chapters of this book describe what is known about how the three pounds of bees we call a honeybee swarm makes a decision about where it will build its new home. The starting point was the mystery of how a bunch of tiny-brained bees, hanging from a tree branch, can make a good choice for their future living quarters and can take timely action on their decision. We then reviewed the observational and experimental evidence concerning each specific mechanism of the house-hunting process—an ingenious and sophisticated tangle of behaviors, communication systems, and feedback loops. Throughout, we've seen that a swarm of bees is a democratic decision-making body that is remarkably accessible to analysis, since we can easily observe what the individual bees are doing while an entire swarm is conducting its decision-making process. It is an amazing bit of good fortune that all of the important individual-level actions occur in full view, either on the surface of the swarm or at the prospective homesites, not deep inside the mass of tightly clustered bees. It has been important to first work our way through the nuts and bolts of the bees' house-hunting process, so we really know this process, but now it is time to step back from the detailed analysis and synthesize what we know by considering the general features of a swarm as a decision-making system.

In doing so, we will find it useful to compare what is known about the mechanisms of decision making in bee swarms and primate brains. This may seem a bizarre comparison, for swarms and brains are vastly different biological systems whose subunits—bees and neurons—differ greatly. But these systems are also fundamentally similar in that both are cognitive entities that have been shaped by natural selection to be skilled at acquiring and processing information to make decisions. Furthermore, both are democratic systems of decision making, that is, ones in which there is no central decider who possesses synoptic knowledge or exceptional intelligence and directs everyone else to the best course of action. Instead, in both swarms and brains, the decision-making process is broadly diffused among an ensemble of relatively simple information-processing units, each of which possesses only a tiny fraction of the total pool of information used to make a collective judgment. We will see that natural selection has organized honeybee swarms and primate brains in intriguingly similar ways to build a first-rate decision-making group from a collection of rather poorly informed and cognitively limited individuals. These similarities point to general principles for building a sophisticated cognitive unit out of far simpler parts.

Conceptual Framework for Decision Making

In essence, decision making is a process in which information is acquired and processed in order to make a choice between two or more alternatives. Thus a honeybee swarm performs decision making when it obtains information about the qualities of a dozen or more potential homesites, processes this information, and selects the most desirable site for its new residence. A good example of a primate brain performing a decision-making task is when a monkey is presented with a visual display consisting of a cloud of white dots moving against a black background (fig. 9.1). Most of the dots move randomly, but a small fraction move coherently in one of two possible directions, left or right. The monkey has been trained to decide whether the motion direction of the coherently moving dots is left or right and to indicate its decision by making an eye movement to a left

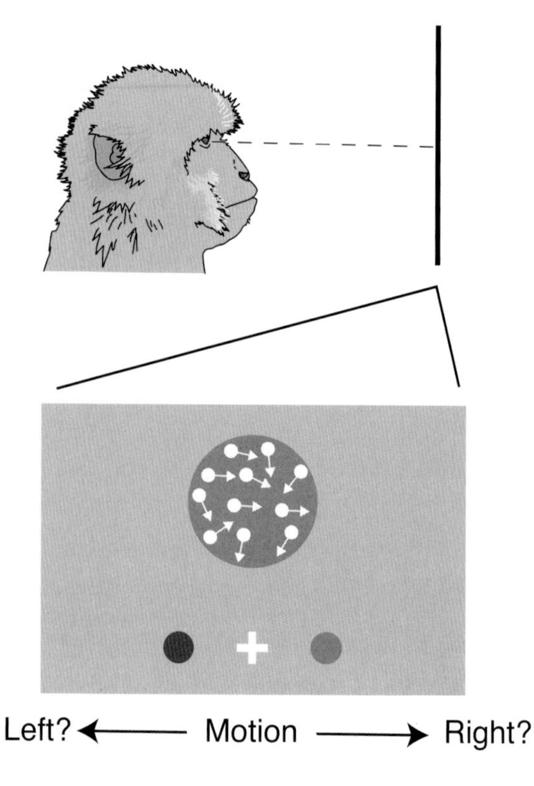

Fig. 9.1 The layout of a perceptual discrimination task in which a monkey indicates its judgment regarding the direction of coherent motion in a moving dot display with a left or right eye movement from a fixation point (white cross) to one of two targets (gray circles) that are aligned with the axis of the coherent motion.

or right target. The percentage of the displayed dots that move coherently can be varied to make the quality of the information higher or lower, and thus the decision-making task easier or harder.

While other behavioral biologists and I have been working to understand the mechanisms of a honeybee swarm's decision making at the level of individual bees, neuroscientists have been working to understand the mechanisms of a human brain's decision making at the level of individual cells. The best progress in unraveling the neural basis of human decision making has been made by studying monkeys (serving as human surrogates) as they perform the decision-making task just described in which a monkey sees a "noisy" visual stimulus, makes a decision in the two-alternative (left or right) choice test, and indicates its decision with an eye movement. By recording neural activity in the various areas of the brain involved in reporting visual information, in processing this information, and in controlling eye movements, neurophysiologists have identified the neural processes that underlie this particular decision-making task.

The starting point is the middle temporal (MT) area of the brain, which processes sensory information about the motions that the monkey sees (fig. 9.2a, b). Each neuron in the MT area has a receptive field that corresponds to a particular portion of a monkey's entire visual field. Also, each MT neuron is motion sensitive for a particular direction of movement, that is, it is activated to fire when a stimulus moves across its receptive field in the preferred direction, and it is inhibited from firing when the moving stimulus travels in the opposite direction. Thus the population of neurons in the MT area is a set of direction-tuned motion detectors that report, in their firing rates, information about the strength of visual movement in their preferred directions within their particular portions of the monkey's visual field. Collectively, they provide the monkey's brain with information regarding the strength of rightward motion and leftward motion over the monkey's whole visual field, hence over the full display of moving dots. At any given instant, however, this information is somewhat ambiguous because of randomness in the moving dot display and because of noise (random fluctuations) in the representation of the information by the MT neurons.

The next step in the monkey's decision-making process occurs in the lateral intraparietal (LIP) area of the brain. The neurons in this area receive inputs from the MT area, and they are organized into direction-specific integrators that sum over time the noisy information being provided by the corresponding MT neurons (fig. 9.2a, b). Thus, as time progresses in the decision-making task, evidence about what the monkey is seeing accumulates in the LIP neurons. For example, if a monkey is watching a visual display containing rightward-moving dots, then LIP neurons that function as rightward-motion integrators will increase gradually their firing rates. The rate at which their firing rates increase depends on the stimulus strength, that is, the number of the dots moving to the right. Also, the various integrators corresponding to different motion directions are mutually inhibitory. One effect of this mutual inhibition is that even if the firing rates of the LIP neurons associated with rightward and leftward motion increase at approximately the same rate at first, later only those neurons associated with the stronger stimulus (rightward motion) will continue to

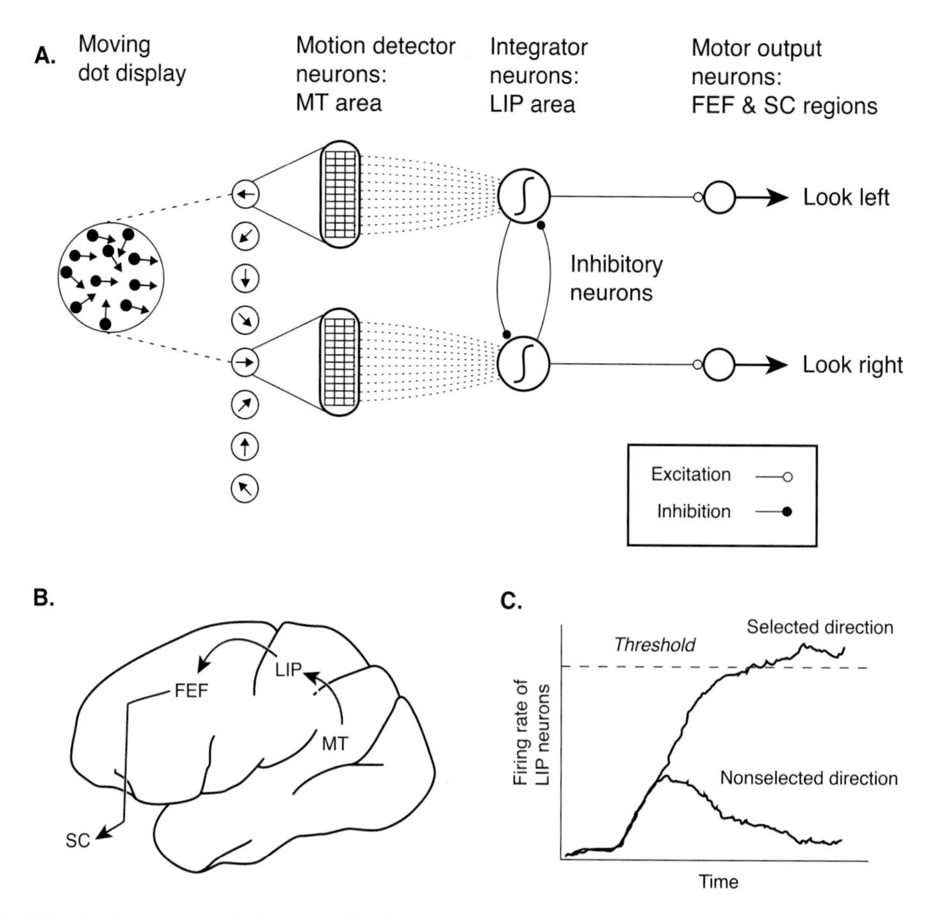

Fig. 9.2 A: Summary of the neurobiological processes underlying the decision of eye-movement direction based on a perception of the motion direction of a visual stimulus. Pools of neurons in area MT extract from the moving dot display the instantaneous strength of visual motion in each direction. The instantaneous pooled estimates of motion strength in each of the two possible directions are passed to integrator neurons in the LIP area that sum over time the inputs from the MT neurons to produce an estimate of the average motion strength in each direction. The integrator neurons corresponding to different motion directions are mutually inhibitory, and they project to neurons that produce eye movements. B: Some of the sites in the primate brain from which decision-related neural activity has been recorded, including the middle temporal area (MT), lateral intraparietal area (LIP), frontal eye field (FEF), and superior colliculus (SC). C: Neural activity associated with the decision transformation has been recorded in the LIP area. Early activity does not discriminate between the two directions, but later the neurons associated with the selected direction show increased firing while those associated with the nonselected direction show decreased firing.

increase their firing rate; those associated with the weaker stimulus (leftward motion) will start to decrease their firing rate (fig. 9.2c). Each population of LIP neurons inhibits the others to a degree proportional to its level of activity, so eventually only the neurons in one LIP population will have high firing rates. This mutual inhibition improves the monkey's discrimination by enhancing the perceived difference in strength between rightward and leftward stimuli, and so helps the monkey avoid attempting to make rightward and leftward eye movements simultaneously.

When the activity of one integrator exceeds a threshold, the decision is made and an eye movement in the appropriate direction is initiated. The eye movement is driven by the output (motor) neurons in the final stage in the monkey's decision circuit. These are neurons in the frontal eye field (FEF) and superior colliculus (SC) regions of the brain, and they receive inputs from the LIP area. Here again, the FEF and SC neurons are direction-specific; each neuron will drive an eye movement in just one direction.

Leo Sugrue, Greg Corrado, and William Newsome, neuroscientists at Stanford University, have devised a helpful conceptual framework for thinking about the multiple stages of information processing that underlie making a simple perceptual decision, like what we've been considering (fig. 9.3). Their framework contains three stages or transformations. First, a sensory transformation converts the information about the external world that has been registered by the animal's sensory organs into a "sensory representation," which makes the information available for further processing within the animal's brain. This is what the MT neurons do in the monkey's motion-detection task. Second, a decision transformation converts the sensory representation into a set of probabilities for adopting the alternative courses of action. In the monkey's brain, this transformation is implemented by the LIP neurons, as they convert the sensory representation of visual motion into a set of "evidence accumulations," specifically the set of firing rates of the integrators representing different motion directions. The level of firing in a particular integrator population determines the animal's relative probability of choosing the alternative represented by this population. Third, an "action transformation" converts this set of probabilities into a specific behavioral act. This final process of action implementation is performed in the monkey's brain

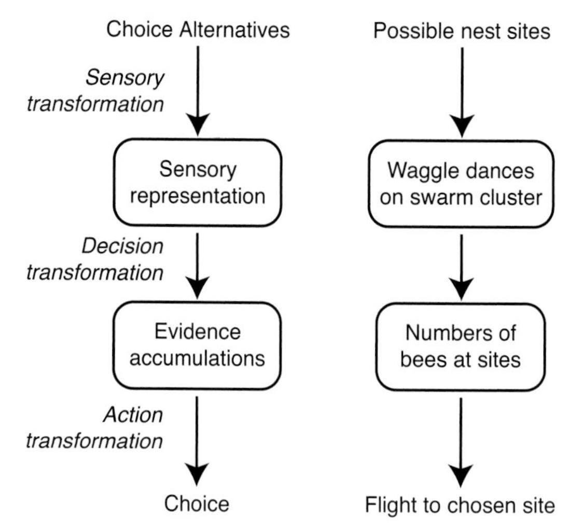

Fig. 9.3 Conceptual framework for decision making that illustrates the processing stages for making a decision (left) and the application of this framework to the mechanisms of nest-site choice by a honeybee swarm (right).

by motor output neurons in the FEF and SC regions when they are activated by the population of LIP neurons whose firing rates have reached a threshold level.

Remarkably, even though this conceptual framework was devised for helping us understand decision making in primate (including human) brains, it can also help us conceptualize the decision-making process in honeybee swarms. In both types of decision-making systems, sensory units create a representation of the outside world inside the system. Also, in both types of systems the processing of the information in the sensory representation consists of a competition between mutually inhibitory integrators of the information (evidence) flowing into the system. And finally, in both brains and swarms, the decision is made when the accumulation of evidence in one integrator reaches a sufficiently high (threshold) level.

The Sensory Transformation in a Swarm

In reflecting on the structural parallels between swarms and brains, I like to think of a swarm as a kind of exposed brain that hangs quietly from a tree branch but is able to "see" many potential nest sites spread over a vast expanse of the surrounding countryside. As we have seen, what gives a swarm such an immense "visual field" is its squadron of several hundred scout bees who fly out for several

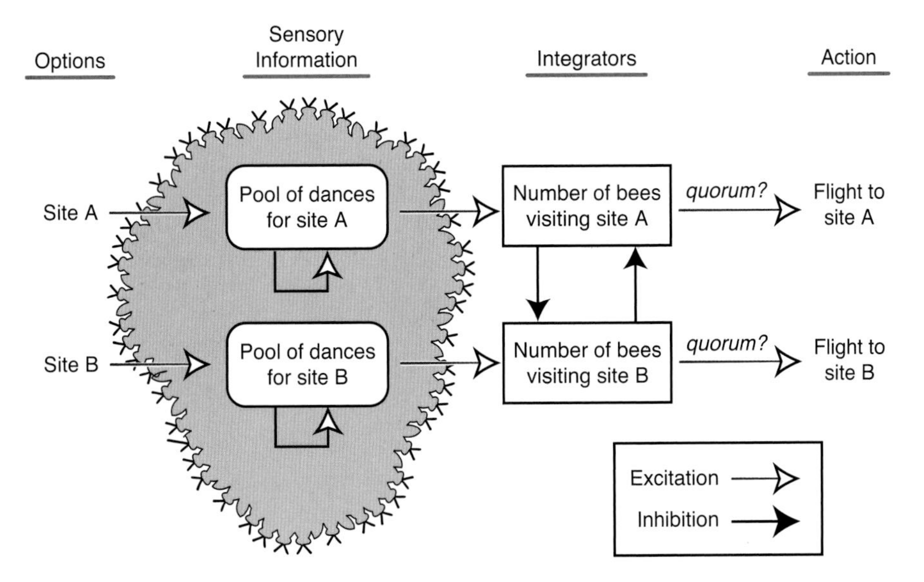

Fig. 9.4 Summary of the behavioral processes in honeybee swarms that produce a nest-site choice based on the qualities of potential homesites. Separate groups of scout bees scrutinize each site and report their estimates of site quality in a pool of dances on the swarm cluster. Because each dancing bee recruits additional bees to dance for her site, there is self-amplification (positive feedback) in this reporting process. The instantaneous pooled estimates of quality for each site activate additional bees to visit the sites. The number of bees visiting each site integrates over several hours the instantaneous pooled estimates of each site's quality and provides an estimate of the relative quality of each site. The numbers of bees at each site are mutually inhibitory. When the number of bees reaches a quorum (threshold) level at one site, the swarm flies to this site.

kilometers in all directions and scour the environment for prospective dwelling places. We now know that when a scout bee finds a site sufficiently desirable to be worthy of attention by others, she flies back to the swarm and reports her find on the swarm's surface by performing a waggle dance. We also now know that the strength of her dance—the number of dance circuits performed—is proportional to the quality of the site. Thus a scout bee functions as a sensory unit of the swarm, one that transduces the quality of a nest site into the strength of a dance signal. It should be noted too that each scout is a site-specific sensory unit, for each bee reports on just one nest site in the surrounding area, much like each MT neuron reports on just one small portion of the visual field. Over time, as dozens of scout bees return to the swarm and perform dances, they gradually deliver a body of sensory information about the locations and qualities of the potential nest sites that they have found (fig. 9.4). This display of bee dances can be thought of as the swarm's sensory representation of the landscape of possible nest sites. It is analogous to the pattern of MT neuron firings that forms a monkey's sensory representation of the stimuli moving across its visual field.

Several features of the way that the scout bees build their swarm's sensory representation are worth noting, for each makes an important contribution to the success of a swarm as a decision-making system.

1. *The sensory apparatus of a swarm is a sizable population of scout bees.* By fielding several hundred scout bees, a swarm is able to gather a wealth of information about potential nest sites, usually within just a few hours. We have seen, for example in figure 4.7, how a swarm's scouts can locate, inspect, and report on nearly a dozen possible home sites in an afternoon. Also, by distributing the information-collection process among numerous bees, a swarm averages out the bee-to-bee variations in strength of dancing for the sites and thereby increases the accuracy of its information acquisition.

2. *Scouts collect sensory information for several hours or several days.* It is important that swarms base their decisions on an extended sequence of samplings of sensory information because this information is acquired sporadically, especially at first. We have seen that even with several hundred scouts exploring simultaneously, several hours may pass before one of them returns with news of an outstanding find. And even after news has been received about all the alternatives, the further reporting on them tends to be episodic, as is shown in figure 6.5. A lengthy period of information gathering enables a swarm to assemble a sizable, and thus reliable, body of sensory information on each site.

3. *Each scout makes an independent evaluation of a site.* Even though most of the scouts reporting on a site are recruited to it, the recruitment process only brings a scout to a site. It does not compel her to report favorably on the site. Instead, each scout makes an independent evaluation and decides for herself how strongly to announce the site when she returns to the swarm and performs a dance. This independence of the scouts means that an evaluation error made by one bee won't

be propagated or amplified by blind imitation, and this helps ensure that the total amount of dancing for a site in a swarm's sensory representation is an accurate indication of the quality of the site.

4. *Scouts reporting on a site recruit additional scouts to the site*. Recruitment by scouts creates positive feedback in the number of scouts reporting on a site as recruited bees become recruiters. This means that the sensory information representing a particular site can amplify itself. Because the strength of each scout's dance depends on her site's quality, the positive feedback (amplification) is stronger for higher-quality sites, and eventually the better sites will tend to monopolize the display of dances on a swarm. Thus over time, a swarm's sensory input, or attention, will become focused on superior sites (see figs. 4.2, 4.3, 4.6, and 4.7).

5. *Scouts reduce their dance responses over time*. Even though the quality of a nest site generally does not change, the dances produced by each scout reporting on it gradually weaken over time (see figs. 6.9, 6.10, and 6.11). This decay in the dance response gradually purges a swarm's sensory representation of information about inferior sites. The purging occurs because scouts that report on poor sites with weak dances tend not to attract replacements, so the feeble reporting of these poor sites withers away. Thus the decay in the dance response also contributes to the way that a swarm, over time, increasingly focuses its attention on better sites.

6. *Scouts may adaptively choose between exploring versus exploiting*. It remains to be shown, but it may be that scouts choose between exploring for unknown (and potentially better) sites versus exploiting already known sites, and that they do so by sensing the abundance of dances on the swarm. If so, then this would endow a swarm with a means of regulating its intake of sensory information, increasing it when the swarm's sensory representation is still poorly formed, and limiting it when the swarm is well supplied with sensory information.

Besides these six features of scout bees as sensory units that foster successful swarm decision making, there are two features that almost certainly *hamper* successful decision making. The first is that scouts often make their reports asynchronously, which means that at any given moment their dances will tend to provide a poor indication of the true qualities of the alternative sites. For example, in figure 6.5, we can see that from 10:00 to 10:15 all of the dancing on

faster and become a larger fraction of the dances on the swarm (see fig. 6.7). One can think of the mutual inhibition among integrators as a means of preventing the emptier ones from refilling after leaking.

Indeed, another shared design feature of the integrators in monkey brains and honeybee swarms is that they are leaky. In other words, in both systems, the accumulation of evidence in any given integrator declines unless additional evidence flows into it. In chapter 6, we saw how each scout bee's commitment to advertising and visiting "her" site steadily declines over repeated visits to the site (figs. 6.5 and 6.9), hence each scout eventually leaks from the accumulated evidence supporting the choice of her site. Leakage in the accumulation of evidence is a key feature of several models developed by mathematical psychologists to model the information processing that underlies decision making in primate brains (e.g., the "leaky, competing accumulator model" developed by Marius Usher of the University of London and James McClelland of Stanford University). In these models, leakage evidently improves decision making by increasing the time over which the noisy evidence accumulates until sufficient information for a decision is gained. Leakage also enables a decision-making system to update itself if the situation changes, as when a superior alternative is discovered. In other words, leaky integrators help a system avoid producing fast mistakes.

This explanation for the function of leakage evidently applies also to honeybee swarms. I make this claim based on what Kevin Passino and I learned when we explored the design of a swarm as a decision-making system by building a mathematical model of the nest-site selection process. Our model simulated the activities of 100 scout bees presented with a landscape containing six nest sites that differed in quality. Each scout bee was endowed with all of the known behavioral rules of these bees: uncommitted scouts search for new sites or follow dances to be recruited to known sites, committed scouts evaluate their sites and advertise them with dances whose strengths depend on site quality, and so forth. We first checked the validity of our model by testing whether it would replicate real-world examples of nest-site selection, like those represented in figure 5.7. In fact, it does so beautifully. Then we used our model to create "pseudomutant" swarms—ones whose scout bees behave a bit differently from what we see in nature—that would show us how small changes in the behavioral rules of the scouts affect a swarm's decision-making performance. For example, we varied the dance decay rate of scout bees to see how this affects the speed and accuracy of decision making by swarms. In nature, the average scout bee reduces the strength of her dancing by 15 dance circuits per trip back to the swarm (see figs. 6.10 and 6.11), so we looked at what would happen if this dance decay rate were raised (up to 35 dance circuits per trip) or lowered (down to five dance circuits per trip). Changing the dance decay rate also changes the integrator leakage rate, since a scout bee stops visiting a site—hence leaks from the integrator—shortly after she stops dancing for the site.

We found that when we lowered the dance decay rate, so that bees continued dancing longer and "leaked" from the nest site more slowly, our model swarms made more rapid but less accurate decisions. Their decision making deteriorated because slowing the leakage accelerated the evidence accumulation at all the sites, so if the best site happened to get discovered late, one of the inferior sites could accumulate the threshold level of evidence first and win the competition among the sites. Conversely, when we raised the leakage rate, our model swarms made less rapid but more accurate decisions. They were sluggish decision makers because the scouts quit visiting their sites so quickly that even the best site's integrator had difficulty accumulating the threshold level of evidence. It was extremely pleasing to discover that the dance decay/scout leakage rate that we measured in natural swarms is such that these swarms operate with a good balance between speed and accuracy in choosing their homes.

The Action Transformation in a Swarm

The final stage of the information processing that underlies decision making is the rendering of a single response from the multiple readouts of all the integrators. It is now clear that both in monkey brains making eye-movement decisions and in honeybee swarms making nest-site choices, a response is made when the evidence accumulation in one of the integrators reaches a threshold level. In both systems the mechanism for choosing a discrete response from a distribution of integrator states is simply one of letting the choice fall to whichever alternative first gains the threshold level of evidence in its integrator. This usually generates a good decision because the relative level of evidence in each alternative's integrator normally reflects the relative strength or quality of each alternative. We have seen, for example, how the better the candidate nest site, the stronger the dances produced to report it, and the swifter the buildup of scout bees at the site. Moreover, the self amplification of the sensory input for each alternative (as recruits become recruiters) and the mutual inhibition among the integrators for the alternatives (by competition for the uncommitted scouts) help ensure that the best nest site will prevail in the contest to accumulate the critical level of evidence, even if the best candidate enters the contest late, as often happens (see figs. 4.7 and 5.7).

We have seen in chapter 7 that the decision-making system of a honeybee swarm senses when one of the alternatives has amassed the threshold level of evidence by means of quorum sensing. That is, the scouts at each candidate site somehow monitor how many of them are at the site, and they note when they have assembled the threshold number (quorum) needed to take action. We have also seen that when the scouts at the chosen site have sensed a quorum they stimulate the swarm to prepare to take action by returning to the swarm and producing worker piping signals that stimulate the nonscouts to warm up their flight muscles. It seems likely that the worker piping signals also stimulate any scout still committed to a losing site to quit this site. This way, while the nonscouts in the swarm are preparing for flight, the scouts are consolidating the consensus they must build lest they give mixed guidance signals when the swarm takes flight. Eventually, once all the bees in the swarm cluster have warmed their flight muscles to a flight-ready temperature of 35+°C, the scouts who primed the swarm for flight with piping signals begin to trigger the swarm into flight with buzz-running signals (see fig. 7.13). Finally, the scouts who know the way to the chosen site steer the swarm along its chosen course of action.

A critical element in the design of this decision-making system is the quorum size, for it turns out that it strongly influences the speed and accuracy of a swarm's choice of its new home. This fact was revealed when Kevin Passino and I turned up and down the quorum number of bees in our mathematical model of the bees' nest-site selection process. We found that adjusting the number downward from its normal value—some 15 bees present simultaneously outside the nest site—caused swarms to make quick but error-prone decisions, while adjusting it upward gave rise to slower but only slightly more accurate decisions. It looks, therefore, like the bees normally operate with a quorum set high enough to guarantee that swarms make highly accurate decisions rather than super speedy ones. This makes sense, for a swarm has just one crack at correctly making the life-or-death choice of its dwelling place, so it should choose carefully, not rapidly. The high quorum number may also be favored by a swarm's need to have a sizable crew of scouts who have visited the chosen site and so can guide the swarm to its new residence. There does exist the possibility that the bees will lower the quorum number in an emergency, such as when the weather turns dangerous or the swarm begins to starve. This way, a swarm that is in mortal danger may gain some shelter without further delay. Whether this possibility is an actuality remains, however, a subject for future study.

Convergence on Optimal Design?

Thirty years ago, in his book *Gödel, Escher, Bach: An Eternal Golden Braid*, the computer scientist Douglas Hofstadter presented the intriguing idea that "ant colonies are no different from brains in many respects." He pointed out that in both systems a higher-level intelligence emerges from groups of "dumb" beings: groups of ants behaving and groups of neurons firing. At the time of Hofstadter's book, the similarities between social decision-making systems and neural decision-making systems could be seen only hazily, for example, by noting that both kinds of systems encode information about the external world in the activity patterns of their elements. Now we know a great deal more about the decision-making mechanisms of insect societies and primate brains, and what we have learned over the past three decades provides striking support for Hofstadter's idea that evolution has built intellectual strength in ant (or bee) colonies and in primate brains using fundamentally similar schemes of information processing.

We have recently realized that primate brains and honeybee swarms are faced with the same basic problem of choosing between alternative courses of action based on a body of noisy information that is dispersed across many component

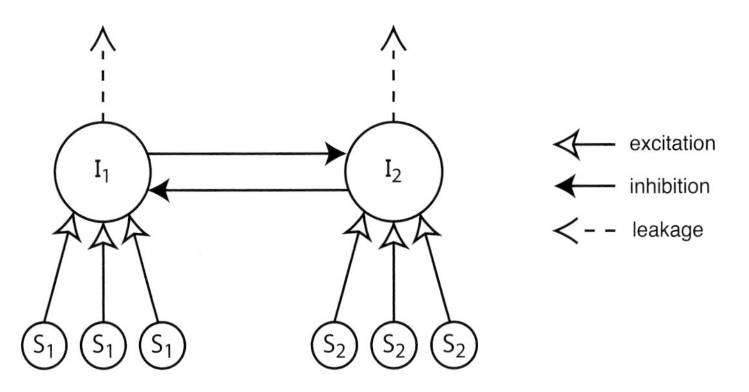

Fig. 9.5 A model of decision making in primate brains and honeybee swarms. In each, populations of neurons or bees represent accumulated evidence for the alternative choices. These populations (I_1 and I_2) integrate noisy inputs from sensory units (S_1 and S_2), and they slowly leak their accumulated evidence. Each population also inhibits the other in proportion to its level of activity (neurons) or its size (bees).

parts, none of which will ever acquire global knowledge of the alternatives. And as we have seen, the solution that both have hit upon is an information-processing system that has the design shown in figure 9.5. This design has five critical elements:

1. A population of sensory units (S_i) that provides input about the alternatives. Each sensor reports (noisily) on just one alternative, and each sensor's input strength is proportional to the quality of its alternative.
2. A population of integrator units (I_i) that integrate the sensory information over time and over sensory units. Each integrator accumulates evidence in support of just one alternative.
3. Mutual inhibition among the integrators, so the growth in evidence in one suppresses with increasing strength the growth of evidence in the others.
4. Leakage of the integrators, so the growth of evidence in an integrator requires sustained input of sensory evidence supporting its alternative.
5. Threshold sensing by the integrators, such that the decision falls to the alternative whose integrator first accumulates a threshold level of evidence.

What underlies this striking convergence in the design of decision-making systems built of neurons and bees? (Also of ants; a beautiful set of studies of collective decision making during house hunting by the rock ant *Temnothorax albipennis* has revealed an information-processing scheme that is remarkably similar to the one described here for honeybees, though of independent evolutionary origin.) A strong possibility is that this striking similarity exists because this design is a means of implementing robust, efficient, and possibly even optimal decision making. It has been shown mathematically that the scheme shown in figure 9.5 can implement the statistically optimal strategy for choosing between two alternatives. This is the sequential probability ratio test (SPRT), which specifies when to stop integrating further evidence in order to achieve a given error rate. Among all possible tests, this one minimizes the decision time for any desired level of decision accuracy. In other words, this test achieves the optimal trade-off between decision accuracy and decision speed.

Recently, James Marshall, a computer scientist at the University of Bristol in England, and a team of colleagues, have examined theoretically how honeybee swarms might implement optimal decision making in the simple situation of a binary choice between two possible homesites. They point out that in a race between two evidence totals, the evidence for one alternative can be seen as evidence against the other, so in effect the evidence can be accumulated as a single total. This means that as time passes and the decision-making system acquires evidence for the two alternatives, at any one time only one alternative will have accumulated a nonzero level of evidence in its favor. In other words, the accumulation of evidence can be thought of as a random walk along a time line where the positive direction represents increasing evidence for one of the alternatives and the negative direction represents increasing evidence for the other alternative (fig. 9.6). The drift of the evidence line up or down denotes the tendency of the line to move toward the better alternative, and the jaggedness of the line represents the noisiness or uncertainty in the incoming evidence. It turns out that this random walk or diffusion model of decision making implements the statistically optimal SPRT.

In the case of a swarm making a choice between two possible nest sites, the existence of strong mutual inhibition between the two integrators—the two groups

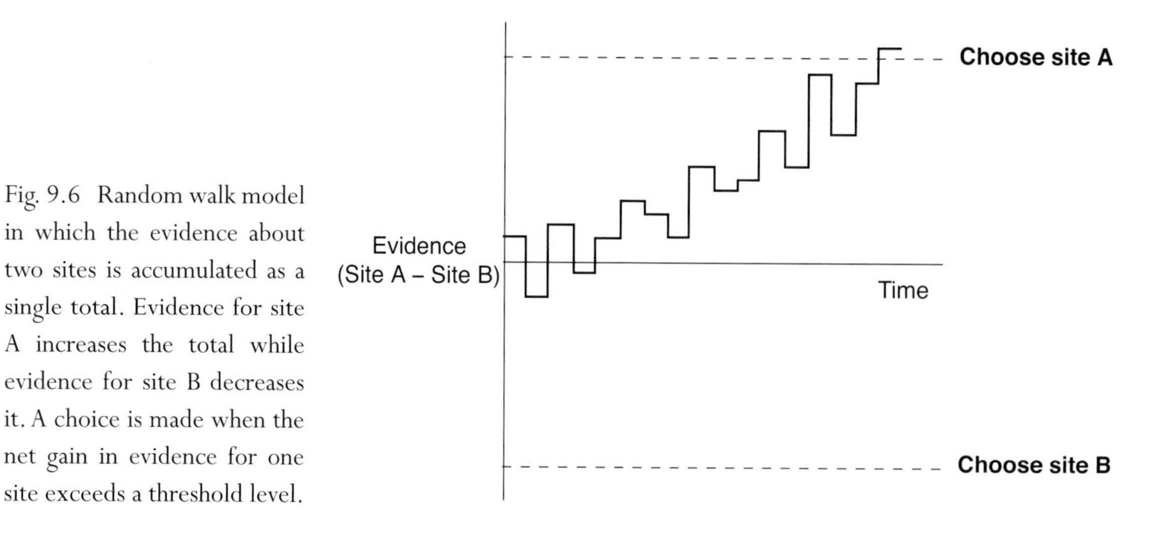

Fig. 9.6 Random walk model in which the evidence about two sites is accumulated as a single total. Evidence for site A increases the total while evidence for site B decreases it. A choice is made when the net gain in evidence for one site exceeds a threshold level.

of scout bees visiting the sites—makes it possible that the evidence for one site will be evidence against the other. Strong mutual inhibition is likely to exist, however, only when there are few uncommitted bees in the swarm, at which time each gain of a supporter by one site will come at the cost of a supporter from the other site. This situation, or at least something close to it, is apt to arise only rather late in the decision-making process, when most of the scout bees have entered the process and have become committed to a site. This is also the time when many of the less desirable sites have been eliminated from the contest and new sites are being discovered only infrequently. Thus it seems that optimal decision making, as modeled by the SPRT, may happen only toward the end of a swarm's decision-making process. But this is probably just when the greatest skill in decision making is needed, for toward the end only a few relatively high-quality sites are likely to remain under consideration, which makes it difficult to identify the best site. Clearly, future studies will need to examine closely whether in a binary-choice situation it is typical for all of a swarm's scouts eventually to become committed to one site or the other, whereupon the decision making between the two sites is expected to proceed optimally.

Of course, in nature, decision makers are rarely faced with the simple binary choice situation, for which the SPRT is provably optimal. Certainly we have seen that most honeybee swarms are faced with choosing among a dozen or more possible nest sites, and that even toward the end of a swarm's deliberations the race to gain a threshold level of evidence often involves more than two sites. Nevertheless, because the SPRT remains effective in situations with several alternatives, so long as some are markedly better than others, it is possible that primate brains and honeybee swarms have independently evolved the same basic decision-making scheme precisely because it provides a good approximation of optimal decision making. If this hunch proves correct, then we are looking at an astonishing convergence in the adaptive design of two physically distinct forms of "thinking machine"—a brain built of neurons and a swarm built of bees.

10

SWARM SMARTS

... for so work the honey-bees,
Creatures that by a rule in nature teach
The act of order to a peopled kingdom.
—*William Shakespeare, Henry V, 1599*

Let us now consider what lessons we humans can learn from honeybees about how to structure a decision-making group so that the knowledge and brainpower of its members is effectively marshaled to produce good collective choices. This is an important subject, for human society relies on groups to be more reliable than individuals when it comes to making weighty decisions. This is why we have juries, boards of trustees, blue-ribbon panels, and nine justices on the U.S. Supreme Court. But as we all know, groups don't always make smart decisions either. Unless a group is properly organized, so that the face-to-face deliberations of its members result in collective reasoning that is broadly informed and deeply thoughtful, the group is apt to be a dysfunctional decision-making body. If so, then the judgments issued by the group can produce fiascoes for the affected community. Fortunately, the house-hunting bees show us a brilliant solution to the puzzle of what gives rise to good group decision making. It is a solution that has been honed by natural selection for many millions of years—fossils from the Oligocene epoch indicate that honeybees have existed for at least 30 million years—so it is certainly a time-tested method for achieving collective wisdom.

Of course, employing insects as management gurus has its limitations, and we should not blindly imitate their methods. Nevertheless, I will claim that the bees demonstrate to us several principles of effective group decision-making and that by implementing them we can raise the reliability of decision making by human groups. The latter part of this claim is not merely hypothetical, for I have applied what I've learned from the bees to humans, especially to my colleagues at Cornell. In 2005, just as the shape of the bees' decision-making process was becoming clear, I became head of the Department of Neurobiology and Behavior. For fun, and as an experiment, I decided to introduce some of the ways the scout bees go about choosing a home to the ways my fellow professors and I hold deliberations in our monthly faculty meetings. Unlike the swarm bees, we do not face life-or-death decisions, but we do face ones that are difficult enough: choices about hirings, promotions, and other matters with long-term consequences for our tight-knit academic community. I am probably blissfully blind to my colleagues' true thoughts about our collective decision making, but I think that they have been satisfied with the tough decisions we've made, even though things certainly haven't always gone the way that each professor wanted. And I'd like to think that their apparent satisfaction reflects how our decisions have been based on open and fair discussions. In any event, I will explain below how I have tried to put the "Five Habits of Highly Effective Groups" that I have learned from the bees to work in a university setting.

To strengthen further my case that what has been learned from the bees has relevance to humans, I will discuss some intriguing resemblances between honeybee swarms and New England town meetings in how they are organized to produce good decisions. Why use the New England town meeting as a point of comparison? It is because this special form of small town government, which has existed for more than three centuries and is arguably the most authentic form of human democracy in the world, uses a collective decision-making process not unlike that used by the swarm bees. Once a year, on Town Meeting Day—traditionally the Tuesday following the first Monday in March—the citizens in a town come together in an open, face-to-face assembly and render binding collective decisions (laws) that govern the actions of everyone in their town. Each town meeting is a fascinating blend of communal ambience and individual enterprise, as is each honeybee swarm. We will see that there are compelling similarities in the inner workings of these two proven forms of democracy. I don't think it is a happenstance that what works well for bee swarms also works well for town meetings.

Lesson 1: Compose the Decision-Making Group of Individuals with Shared Interests and Mutual Respect

For the members of a decision-making group to work together productively, they must have a fair amount of alignment of interests so that they are inclined to form a cooperative and cohesive unit. It is also helpful if the group's members have a fair amount of mutual respect so that they will constructively debate the proposals offered by one another, consider the other individuals' points of view, and refrain from bruising egos and arousing anger when it comes time to critically evaluating one another's ideas. Certainly a decision-making group composed of clashing curmudgeons is unlikely to have the morale and working relations needed to function effectively.

The house-hunting bees exemplify a group whose members have shared interests and mutual respect. Biologists now understand that the genetic success of each worker bee in a honeybee colony depends on the fate of the entire colony; no individual bee succeeds unless the whole colony survives and reproduces. Furthermore, it is now understood that the virtual absence of reproduction by the worker bees in a honeybee colony means that these bees all propagate their genes through one shared channel: the reproductive offspring of their mother queen. And since these reproductive offspring—the queens and drones produced in the spring—contain an unbiased sample of the colony's genes, the colony propagates the workers' genes with a high degree of fairness. So, because the workers have a common need for their colony to thrive, and because a thriving colony passes the workers' genes into the future with near perfect impartiality, it is not surprising that the workers of a honeybee colony cooperate strongly to serve the common good.

The humans in a community rarely share a singularity of purpose like the bees in a swarm, so humans are less inclined than bees to be highly cooperative when tackling a problem they must address together. Nevertheless, there are certain things we can do to encourage ourselves to work together. One is for a group's leader to remind the members at the outset that they all have a large stake in the welfare of the group. At the start of the annual town meeting in Bradford, Vermont, for example, the moderator—Larry Coffin, who has served in this role for 38 years and probably knows more about this job than anyone else in the state—begins the meeting in the traditional way by asking for a moment of silence "out of respect for the exercise in democracy that we are about to engage in." This gently reminds everyone in the auditorium that they have assembled to make decisions and pass laws for *their community*. Similarly, at the start of the monthly faculty meetings in my department at Cornell, I generally make a few remarks reminding everyone that our overarching goal is to make decisions that will strengthen our department and so ultimately benefit us all.

A second way to foster good working relations within a human group charged with a decision-making task is to stock it with genuinely reasonable people, ones who are known to be respectful of others and constructive in their comments while at the same time good at spotting hidden problems and engaging in vigorous debate. Often, however, one cannot choose the members of a decision-making group. But even when the personality mix of a working group cannot be shaped, norms of behavior and procedural practices that foster good morale can be promoted. In Bradford, Vermont, for instance, Larry Coffin reminds everyone at the start of the yearly town meeting to address their comments and opinions directly to him, the moderator, rather than to other citizens. This helps keep tempers from flaring and the debate moving forward. Likewise, in my department's faculty meetings, I sometimes find myself gently cutting off demoralizing stalemates by noting when contrary viewpoints are being needlessly repeated. And twice I have had to cool an overheated exchange between two faculty members by nudging them off a personal quarrel. Such things reawaken my appreciation of the marvelous absence of corrosive relations among the debating bees.

Lesson 2: Minimize the Leader's Influence on the Group's Thinking

One of the most striking features of the swarm bees' decision-making process is that it is a perfectly democratic endeavor, one in which the power is evenly diffused among all the scout bees in a swarm. In other words, the swarm bees choose their new home without a leader integrating information from different sources or telling the others what to do. Even the all-important queen, who is certainly the genetic heart of a swarm, is merely a bystander. Indeed, in many of the experiments described earlier in this book the swarm's queen was confined in a small cage (around which the swarming bees clustered), so she was physically separated from the scout bees' deliberations, and yet the swarm skillfully chose its new home. By operating without a leader, the scout bees of a swarm neatly avoid one of the greatest threats to good decision making by groups: a domineering leader. Such an individual reduces a group's collective power to uncover a diverse set of possible solutions to a problem, to critically appraise these possibilities, and to winnow out all but the best one.

Unlike honeybee swarms, most human groups operate with a leader. So clearly, a prominent question we must address is how the leader of a decision-making body should behave to promote sound thinking by the group. I suggest the answer is that the group's leader should act as impartially as possible, so that his or her influence on the outcome of the decision-making process is minimized. Only then can the group fully exploit the power of collective choice. This means that at the start of deliberations the leader should limit his or her comments to neutral information about such things as the scope of the problem, the resources available to solve it, and the rules of procedure. Also, the leader should refrain from advocating any solutions he or she would like to see adopted and instead should show an open-minded desire for fresh ideas. By functioning not as a proselytizing boss but as an impartial information seeker, the leader creates an atmosphere of open inquiry that helps the group tap its summed knowledge to assemble a wide range of possible actions. In addition to conducting meetings in a nondirective way, the leader should encourage the airing of doubts and disagreements, even ones that are critical of the leader. This fosters the free discussion and careful debate that the group will need to thoroughly evaluate its options.

If a leader shows partiality at the outset of the deliberations, or expresses displeasure if the discussion is not going in a certain direction, then he or she is likely to subvert good group decision making. One problem with both of these leadership practices is that they can lead to a premature consensus by the group as its members, consciously or unconsciously, seek to please their leader. An example of this phenomenon is the decision made by President George W. Bush and his foreign policy team to invade Iraq in 2003. As explained by Scott McClellan, deputy White House press secretary at the time, Bush's style of leadership was headstrong. He told his foreign policy advisers of his deeply held belief that Saddam Hussein was an international pariah who possessed weapons of mass destruction and so should be removed. Evidently, Bush's foreign policy advisers, including his national security adviser, Condoleezza Rice, went right along with his thinking as they sought to please the president. They did little to question his thinking, engage in extended debate about the possible policy options, or delve deeply into the consequences of going to war. In short, they squandered their opportunity to use group intelligence. Thus we now know that the hasty and flawed decision to invade Iraq was based largely on the gut feelings of just one man, George W. Bush.

Larry Coffin, the gentleman who has served for nearly four decades as the moderator of town meetings in Bradford, Vermont, shows us how an impartial leader can promote the emergence of a group's collective wisdom. Although the moderator has sole authority for running the town's annual meeting, he or she must always remember that the will of the people comes first. One way that Coffin avoids influencing the townspeople's will is the way he starts the discussion for each question, or Article, on the meeting's published agenda. After reading the Article—for example, "Shall the Town of Bradford purchase a fire truck for an amount not to exceed $306,000?"—Coffin asks the crowd, "What is your pleasure?" Soon a townsperson will raise a hand, Coffin will recognize him or her to speak, and the process of open deliberation on this Article is underway.

In a New England town meeting, the moderator is responsible for making sure that every registered voter is allowed to speak, that the competition among competing views is conducted fairly, and that the group makes its decisions in a timely manner. To fulfill these duties, moderators are instructed not to rely on personal authority, but instead to lean on *Robert's Rules of Order*, which Major Henry M. Robert, an engineer in the U.S. Army, published in 1876 as a "guide to fair and orderly procedures in meetings." If a town's moderator defers to these rules, and so acts with some humility, then the general will of the town's citizens will emerge on each problem they address.

Lesson 3: Seek Diverse Solutions to the Problem

Sometimes a problem's architecture defines the possible solutions—to open a door, we know our choices are limited to pushing or pulling—but other times the available options are not well defined. Then the logical first step toward solving the problem is to uncover a profusion of possible solutions in the hope that one will prove excellent. And here is where a democratic group can vastly outperform a despotic individual, since a group's power to explore for options can greatly surpass that of a lone individual. This is especially true if the group's members are numerous, diverse, and independent. With many individuals bringing unique experiences to the problem and searching independently for possible solutions, the chances are high that someone will come up with a radically new option, which might be just what is needed.

The house-hunting bees provide a beautiful demonstration of the effectiveness of a large and diverse search committee whose members explore on their own. As we have seen, a swarm sends out hundreds of scout bees that explore for potential homesites over an area stretching five or more kilometers (at least three miles) from the bivouac site. Each intrepid scout bee works by herself, diligently poking around tree trunks and rock outcrops in search of small, dark openings that might lead to a suitably roomy and protective nest cavity. Whenever a scout chances upon a possible dwelling place, she scrutinizes it and, if it proves acceptable, she returns to the swarm and freely reports her discovery with a waggle dance. This puts an option on the table for further consideration. It is as if each scout that announces a new site says to her fellow scouts, "Shouldn't we give some thought to this possibility, which is located X degrees to the right (or left) of the sun and Y meters away?" The distributed reconnaissance process of the scout bees often continues for hours or days, so it is not surprising that a swarm typically uncovers 10 to 20 or even more possible places to live. Clearly, the house-hunting process of a honeybee swarm is open to the widest possible array of choices, and this gives the bees a strong start in selecting the best available living quarters.

What can we humans do so that our own decision-making groups, when faced with complex problems, also develop a broad set of alternatives from which to choose? Considering what the bees do, I suggest four things. First, make sure the group is sufficiently large for the challenge it faces. Second, make sure the group consists of people with diverse backgrounds and perspectives. Third, foster independent exploratory work by the group's members. And fourth, create a social environment in which the group's members feel comfortable about proposing solutions. If a group implements all four suggestions, then it is likely to achieve a thorough exploration of its options.

Often one cannot shape all of these four elements of a group's search for alternative solutions, but it will still help to improve some of them. For example, in organizing a faculty meeting in my department at Cornell, I cannot adjust its size or composition. I can, however, encourage creative thinking about possible solutions and can foster the reporting of them to the group. To help get new ideas, I will present the problem to my colleagues well in advance of the meeting. This way, each one can wrestle with it privately prior to the meeting. And to encourage everyone to contribute their ideas, I will begin the meeting by suggesting that we start tackling the problem by getting a wide range of options on the table. My colleagues are always good "scout bees," and most are as uninhibited as a dancing bee about sharing their knowledge, so this brainstorming phase quickly yields a broad set of proposals. But to be sure that we are looking at all the conceivable options, I will ask each person who has not spoken if he or she has something to add. Often, the quieter folks will further broaden the list of options with thoughtful proposals.

Given the importance of endowing a decision-making group with diverse knowledge, it is noteworthy that *Robert's Rules of Order*, which provides the exact rules of procedure for a New England town meeting, includes a nifty rule that helps ensure that every participant at the town meeting gets to express his or her thoughts on each issue; no one may speak more than twice on a particular issue until everyone who wants to speak on the issue has had an opportunity to do so once. So long as a meeting's moderator strictly enforces this rule, nobody can dominate a discussion. This certainly boosts the fairness of a debate. It also boosts the debate's effectiveness, because this rule helps ensure that a town meeting's decision making benefits from the full range of facts and opinions held by its participants. Evidently, Major Henry M. Robert understood the importance to a community's deliberations of tapping fully the collective knowledge of its members.

Lesson 4: Aggregate the Group's Knowledge through Debate

Probably the greatest challenge faced by a group that makes decisions democratically is to know how to turn the knowledge and opinions of its many members into a single choice for the group as a whole. Indeed, this is a problem that has challenged social philosophers and political scientists for centuries. We humans have devised a variety of voting procedures to single out one option from a list of possible choices: majority rule, plurality wins, weighted-voting schemes, and others. However, the problem of social choice is not unique to humans. In many other species, the same problem arises: how should a democratic group's members reach a decision when they strongly disagree?

The house-hunting process of honeybees provides us with an intriguing answer to this question, one that has been shaped by natural selection over millions of years. We have seen that the heart of the bees' decision-making process is a turbulent debate among groups of scout bees supporting different options (potential nest sites). These groups compete to gain additional members from a pool of scout bees who are not yet committed to a site. Whichever group first attracts a quorum of supporters wins the competition. The winning group then goes on to build a consensus among the scouts, so that when it comes time for the scouts to pilot the swarm to its new home, they are in complete agreement about the flight plan.

What is perhaps most impressive about the bees' system of social choice is its ability to distinguish good options from bad ones so that almost always a swarm selects the single best site from among the dozen or more possible homesites that its scout bees have discovered. And what I find most noteworthy about a swarm's skill in decision making is how it arises from a truly ingenious balance between interdependence and independence among the debating scout bees.

The scouts operate interdependently in that they communicate with one another about their swarm's options. This communication is crucial, because it is what makes it possible for one scout bee's news about a dream homesite to percolate among the hundreds of scout bees in a swarm. We have seen how a committed scout can advertise "her" site to uncommitted scouts by performing a waggle dance. The uncommitted scouts who follow another scout's dance are then recruited to the advertised site, and these recruited bees can in turn advertise the site and thereby recruit still more scouts to this particular site. Thus there is potential for runaway growth—positive feedback—in the number of scout bees visiting each site. And we have seen that the better the site, the stronger the dances advertising it, hence the greater the positive feedback for this site. Thus, by grading their dance advertisements according to site quality, the scouts adaptively bias the competition for more supporters in favor of the superior sites. And once a bias in favor of the better sites is established, it will grow and grow—the rich will become richer—as the process of positive feedback amplifies the starting bias. Sooner or later the supply of uncommitted scouts will dwindle, further intensifying the competition among groups of scouts and leading ultimately to the bees' interest shooting up at one site while fading away at all the others. Almost always, the best site prevails in this winner-takes-all contest. The system works so well that even when the best site is discovered several hours after all the others, it can still quickly dominate the competition. Such come-from-behind success is made possible by the strongest advertising, and thus strongest positive feedback, coming from the crowd championing the tip-top site.

The interdependence among the communicating scout bees is certainly a crucial part of the social machinery by which they aggregate their many pieces of information about potential homesites. Their capacity for recruitment communication, and the positive feedback it engenders, is what gives their decision-making system the ability to concentrate its attention—the buildup of scouts—on one site. But what guarantees that the scout bees focus on the best site is one small but utterly critical piece of independence among them; each one decides whether to advertise a site, and if so how strongly, based on her own, independent evaluation of the site. No scout bee, not even one that has encountered a wildly exuberant dancer, will blindly follow another scout's opinion by dancing for a site she has not inspected. This is critical. If scout bees were to blindly copy dancers, then their decision-making system would be prone to catastrophic amplifications (again, through positive feedback) of errors in the reports by the first scouts who discover potential homesites. It would be much like what happened in the stock market bubble in the late 1990s, when investors bought stocks in telecommunication and technology companies based on watching what others were buying—the "conventional wisdom"—rather than on checking carefully for themselves the fundamentals of these companies. Mindlessly joining a stampeding herd, investors sunk hundreds of billions of dollars in companies that lacked solid value and eventually went bankrupt.

So, rather than perform *slavish imitations* of dancers, the scout bees perform *judicious imitations*. A scout will copy the dance that informed her of a site, but only after she has scrutinized the site herself and has concluded it truly deserves to be promoted. Thus the scout bees make use of the power of communication to help good ideas spread while at the same time they avoid the risk of creating an information cascade about an inferior site. By evaluating sites independently, they invest their attention wisely.

How can humans use what the bees have demonstrated about aggregating the knowledge and opinions of a group's members to make good choices for the group as a whole? I suggest three things. First, we use the power of an open and fair competition of ideas, in the form of a frank debate, to integrate the information that is dispersed among the group's members. Second, we foster good communication within the debating group, recognizing that this is how valuable information that is uncovered by one member will quickly reach the other members. And third, we recognize that while it is important for a group's members to listen to what everyone else is saying, it is essential that they listen critically, form their own opinions about the options being discussed, and register their views independently.

These three principles will be familiar to residents of New England towns where the annual town meeting is conducted in the old-fashioned way, with citizens coming together on Town Meeting Day and making decisions through face-to-face deliberations. Just as the bees engage in courteous but freewheeling debates, so townspeople hold civil but spirited exchanges of views. Just as the bees share their knowledge and opinions about nest sites with concise dances, so townspeople contribute their facts and feelings about fire trucks, bridge repairs, and tax rates with brief speeches. And just as the bees show support (by dancing and visiting) for sites based on their independent evaluations, so townspeople show support (by shouting aye or nay, standing, or using handwritten ballots) for Articles based on their personal judgments. In both bee swarms and town meetings, the heart of the decision-making process is an open competition of ideas that are publicly shared but privately evaluated.

How do the faculty in my department at Cornell use what has been learned about the bees' means of deliberating to make smart decisions in our meetings? First, just like the scout bees that begin their hard work by searching widely for possible dwelling places, we begin tackling a tough issue by looking broadly at our options (as described above). Next, we use the same method as the scout bees for turning the diverse pieces of information in the minds of many individuals into one chosen course of action: a friendly competition of ideas. To get the give and take of deliberation going, I usually say something like, "Well folks, I'd like us to kick these ideas around a bit." This works. Most of my colleagues are comfortable sharing their thoughts, and those who are quieter will be drawn into the discussion when I go around the room and ask them to share their views. One of the best things about this approach is that it lets different individuals contribute different pieces of the puzzle. One person will mention something that we've overlooked so far about one of the proposals. Somebody else will say he or she doesn't understand the last person's point, and another person will provide clarification. Somebody will then say there is something about one of the proposals that bothers him or her, and others will agree or disagree with this and explain why. If the meeting is going well, there is a noticeable forward movement to the discussion.

Once it feels like everything that needs to be said has been said, and my colleagues indicate that they have the information they need to make a decision, we will take a vote. In the past, we generally took votes by show of hands, but now we use secret ballots. Some people balked at first. "We've never done this before except on tenure decisions!" But after I explained that I really want to get each person's independent opinion, free of peer pressure that could produce conformity, they realized that voting by secret ballot is the best way to know our true collective judgment on an issue.

Lesson 5: Use Quorum Responses for Cohesion, Accuracy, and Speed

One might think that when a democratic group has to make a decision that will apply to everyone in the group, it is best to let the group's debate continue unhindered until the opinions of the participants have coalesced around a unified choice. After all, if a problem has an underlying correct solution, then it might pay to argue things through until everyone accepts this solution. This would both ensure that an accurate decision is made and promote broad acceptance of the decision. Sometimes, however, there isn't one solution that serves everyone's interests, in which case more discussion is unlikely to produce agreement and the best thing is probably to cut off a bitter debate with a vote. But even when there *is* an optimal solution—likely for groups whose members have common interests—it may not be worthwhile arguing on to reach complete agreement. Usually there are costs associated with investing more time in the decision process, and the accumulating costs of further debate can eventually outweigh the benefits.

The house-hunting honeybees show us a clever way for a decision-making group to make an accurate consensus decision and also save some time. Their trick is to have the scout bees make quorum responses, that is, to have these bees make *sharp* changes in their behavior when a threshold number (quorum) of individuals support one of the alternatives. Let's review how this works. We have seen that the bees in a swarm must choose accurately to survive, and that they must stay together to survive, so they need to reach an accurate consensus decision about their new home. We have also seen that these bees will invest heavily, up to several days, in searching for possible homesites and publicly debating which one is the best. And we have seen that once the population of scouts at one of the potential homesites exceeds a threshold, or quorum, the scouts visiting this site will abruptly change their behavior and return to the swarm to perform piping signals. Their piping induces the many thousands of nonscout bees to warm their flight muscles in preparation for the swarm's flight to the chosen site. This piping probably also tells the scouts from the nonchosen sites (the ones without a quorum) that they should cease advertising and visiting these sites, which in turn will speed up the consensus building among the scout bees. Thus, because a quorum of scouts at the winning site triggers key changes in the behavior of the scouts from this site, the decision-making system of honeybee swarms has a means of accelerating consensus formation once evidence sufficient to guarantee an accurate decision has accumulated at one of the sites. Brilliant! An additional benefit of the scout bees' sharp quorum response is that it enables the thousands of nonscouts in a swarm to start their flight preparations long before the scouts reach their consensus, which further shortens the time the swarm spends hanging precariously from a swaying tree branch.

Quorum responses can also help human decision-making groups that need to find agreement do so with high accuracy and all possible speed. For example, in the faculty meetings of my department, when we face a major decision that should be resolved with a unanimous vote one way or the other—such as whether or not to recommend an assistant professor for promotion with tenure—we will take straw polls (by secret ballot) periodically during the discussion to see how close we are to consensus. If a poll reveals that we are far from unanimity, then we all know that further careful debate is needed for everyone to become of one mind. But if a poll reveals that we are close to agreement, the few folks supporting the minority position usually will realize that a collective decision has essentially been reached, that prolonging the debate is pointless, and that it is best to switch to the majority position to build the needed consensus. Thus the device of taking straw polls can give the members of a decision-making group the information they need to make quorum responses that will accelerate their consensus building. Of course, in a human group, as in a bee swarm, individuals should operate with a high threshold when making a quorum response to avoid sacrificing the accuracy of the group's decision making. I believe this is how it works in our faculty meetings. Although I don't know for sure, I estimate that my colleagues will change their votes (and minds?) for the sake of achieving consensus only when at least 80 percent—approximately 16 out of 20—of us are already in agreement. *E pluribus unum* through quorum responses? Yes, but do so carefully, using a quorum that is sufficiently large to ensure accurate decision making by the community.

EPILOGUE

Sixty years ago, Martin Lindauer happened upon a beardlike cluster of honeybees hanging on a bush and noticed something odd: the handful of bees on the swarm that were waggle dancing were black with soot, red with brick dust, and gray with soil. Why were they so grubby? Could it be, he wondered, that while most of the swarm bees had been quietly bivouacked in the bush, these dirty dancers had been out searching for nest sites? With this chance observation, and the insight it sparked, Lindauer embarked on what he would later describe as "the most beautiful experience" of his life: probing the mystery of how a honeybee swarm finds a home.

This book has reviewed how Lindauer and his scientific successors have solved the mystery of how a bunch of bees can wisely choose their new residence. We have seen that this decision is made by a search committee composed of a few hundred scout bees, all of whom have previous experience as foragers but have switched to exploring dark cavities instead of visiting bright blossoms. And we have seen how these house hunters search for candidate dwelling places, share their findings by performing dances, conduct an extended debate about which one is best, and eventually come to an agreement about the swarm's new home. Almost always, the collective wisdom of the scout bees chooses the best available option, so that the swarm occupies a nest cavity that provides good protection and sufficient space to hold the large honey stores that the colony will consume in keeping itself warm throughout winter.

We now know that the amazing feat of democratic decision making performed by the scout bees offers us deep lessons about how a group of individuals *with common interests* can structure their group so that it functions as an effective decision-making body. It is worth taking careful note of how the scout bees manage to be so good at all three of the key ingredients of good decision making by a group: identifying a diverse set of options, sharing freely the information about these options, and aggregating this information to choose the best option.

Remarkably, the scout bees do all these things without working under the guidance of a leader. Doing so certainly steers the bees clear of one of the greatest pitfalls to good group decision making: a dominating leader who advocates a particular outcome and thereby inhibits the group from taking a broad and deep look at its options. But the absence of a leader among the scout bees also means that they operate without the benefit of someone in charge to state the group's objectives, define the group's methods of deciding, keep the group on track during its meeting, foster a balanced discussion among the group's members, and identify when a decision has been reached. The scout bees in a swarm are able to work together well without supervision partly because each bee has a strong incentive to make a good decision; their swarm's survival depends on the scouts finding it a suitably secure and roomy place to live. The success of the leaderless scout bees is also favored by the reality that they have just one problem to solve (so there is no confusion about their objective and no tendency for their discussion to drift off topic) and by the way they have rules of procedure that are hardwired into their nervous systems (so there is no need for someone to define or enforce their rules of procedure). Thus the house-hunting bees remind us that the leader in a democratic group serves mainly to shape the process, not the product, of the group's deliberations. The bees also demonstrate that a democratic group can function perfectly well without a leader if the group's members agree on the problems they face and on the protocol they will use to make their decisions.

The first challenge faced by every decision-making assembly is to identify the available options. Ideally, its members will uncover all the relevant possibilities. We have seen that the house-hunting scout bees approach this ideal by searching widely for prospective nesting sites and discovering a few dozen candidate homesites. The bees' success in finding a broad range of options reflects two things. First, they are a large group, usually several hundred individuals, so they bring considerable bee power to the search for possible places to live. Second, they are a diverse team of explorers, with no two individuals probing the exact same region of the surrounding countryside. For example, one bee will fly off in one direction and examine the dusty knotholes that she finds in the trees on a certain hillside, meanwhile her fellow scouts will set out in various other directions and inspect the cracks in buildings, abandoned woodpecker nests, and whatever other possibilities they encounter. The differences in where the scout bees explore for future accommodations may reflect differences in where they previously worked as foragers, differences in their "personalities" (some may prefer to search far out while others may wish to hunt near by), or differences in combinations of these and other factors. Whatever the exact cause of the variance in where the scouts conduct their reconnaissance, the result is that they discover a broad assortment of possible living quarters. This variety makes it likely that at least one of their finds will provide the bees with an excellent home.

Besides doing a good job of uncovering options, the members of a decision-making group must also do a good job of sharing the news of their finds. If an individual doesn't make the news of her discovery public, but instead keeps it private, this information will go unused, and this can lead to an inferior decision by the group. Imagine, for example, someone in a group uncovering a first-rate option but then not revealing it to others; the group cannot incorporate this information in their discussion. Given the critical importance of exposing all private information that is relevant to a group's decision making, it is not surprising that when a scout bee in a honeybee swarm locates a potential nesting site, scrutinizes it, and concludes that it has high value, she quickly flies back to the swarm cluster and excitedly announces her discovery. We have seen that she does so by performing a waggle dance that reveals the direction, distance, and desirability of her find. The more highly the little scout bee values her property, the more dance circuits she performs, and the more as-yet-uncommitted scout bees she attracts to her site. We have also seen that the scouts who make the original discoveries of potential homesites tend to announce their finds especially persistently, probably to help ensure that the information that they alone (at first) possess gets passed to others and so becomes part of their swarm's pool of public information. It should be noted too that every scout bee is free to advocate whatever site she finds, even one that is a relatively poor option. In a sense, then, on a honeybee swarm, all views are welcomed and respected; all opinions may be voiced.

Once a decision-making group has gathered and shared the information about options, it next faces the challenge of aggregating this information to choose a winner. We have seen that the bees do so in a most ingenious way, by conducting a frank debate among the scout bees supporting the various proposed nest sites. This debate works much like a political election, for there are multiple candidates (nest sites), competing advertisements (waggle dances) for the different candidates, individuals who are committed to one or another candidate (scouts supporting a site), and a pool of neutral voters (scouts not yet committed to a site). Also, the supporters for each site can become apathetic and rejoin the pool of neutral voters. The election's outcome is biased strongly in favor of the best site because this site's supporters will produce the strongest dance advertisements and so will gain converts the most rapidly, and because the best site's supporters will revert to neutral-voter status the most slowly. Ultimately, the bees supporting one of the sites—usually the best one—dominate the competition so completely that every scout bee supports just one site. A unanimous agreement is reached. It is important to note that even though the scout bees' way of making a decision ends with a consensus, the bees do not minimize conflict to reach this consensus. Specifically, there is no suppression of dissenting views in the debate. Moreover, there is no pressure toward social conformity. Instead, each scout bee makes her own, independent decision of whether or not to support a site, based on her own, personal evaluation of the site, not on how others judge the site. Thus the bees aggregate the information about their options by conducting an open debate in which the best site prevails by virtue of its superiority, as judged time and time and time again by dozens, if not hundreds, of independent-minded scout bees.

For millions of years, the scout bees on honeybee swarms have faced the task of selecting proper homes for their colonies. Over this vast stretch of evolutionary time, natural selection has structured these insect search committees so that they make the best possible decisions. Now, at last, we humans have the pleasure of knowing how this ingenious selection process works, and the opportunity to use this knowledge to improve our own lives. Some have said that honeybees are messengers sent by the gods to show us how we ought to live: in sweetness and in beauty and in peacefulness. Whether or not this is true, I believe that the story of house hunting by honeybees can inspire the light of amazement about these beautiful little creatures, a light that I hope has shined through each page of this book.

Notes

Chapter 1. Introduction

Page 3: Quote of George Bernard Shaw, from Shaw, G. B. 1903. *Man and Superman*. Act II, line 79. The University Press, Cambridge, MA.

Page 3: The scope and economic value of honeybee pollination is reviewed in detail in two classic works. McGregor, S. E. 1976. *Insect Pollination of Cultivated Crop Plants.* Agricultural Handbook 496. United States Department of Agriculture, Agricultural Research Service, Washington DC; and Free, J. B. 1993. *Insect Pollination of Crops*. Academic Press, London.

Page 3: Bee taxonomists tell us that there are at least nine species of honeybees, all members of the genus *Apis*. The biology and geographic distribution of each species, including the familiar western honeybee, *Apis mellifera*, are described in detail in Ruttner, F. 1988. *Biogeography and Taxonomy of Honeybees*. Springer-Verlag, Berlin. A recent paper updates what is known about the living and fossil honeybees (genus *Apis*), and discusses the biogeography of these bees given the remarkable discovery recently of a fossil honeybee (*Apis nearctica*) recovered from paper shales of Nevada in the western United States. See Engel, M. S., I. A. Hinojosa-Diaz, and A. Rasnitsyn. 2009. A honey bee from the Miocene of Nevada and the biogeography of *Apis* (Hymenoptera: Apidae: Apini). *Proceedings of the California Academy of Sciences* 60:23–38.

Page 5: A good review of studies on the queen honeybee's pheromonal influence on her workers is provided by Winston, M. L., and K. N. Slessor. 1992. The essence of royalty: honey bee queen pheromone. *American Scientist* 80:374–385.

Page 6: The analogy between the bees in a hive and the cells in a body—in both cases, a constellation of units at one level of biological organization cooperate closely to build a higher-level entity—is discussed in detail in Hölldobler, B., and E. O. Wilson. 2009. *The Superorganism: The Beauty, Elegance, and Strangeness of Insect Societies.* Norton, New York.

Pages 7–8: The study of the neurobiology of primate decision making has made its most striking progress in understanding how simple perceptual decisions are made, that is, how an individual's nervous system transforms sensory information into a perception and then into an appropriate behavioral response. Two recent reviews of this field of study are Gold, J. I., and M. N. Shadlen. 2007. The neural basis of decision making. *Annual Review of Neuroscience* 30:535–574; and Heekeren, H. R., S. Marrett, and L. G. Ungerleider. 2008. The neural systems that mediate human perceptual decision making. *Nature Reviews Neuroscience* 9:467–479.

Page 8: Quote of Henry David Thoreau, from Thoreau, H. D. 1838. Journal entry, March 14.

Page 8: Quote of Friedrich Nietzsche, from Nietzsche, F. 1966. *Beyond Good and Evil*. Random House, New York. Kaufmann, W., trans. 1886. *Jenseits von Gut und Böse*. P. 90. Naumann, Leipzig.

Pages 9–11: A clear description of how Karl von Frisch gradually decoded the waggle dance is found in chapter 3 of von Frisch, K. 1971. *Bees: Their Vision, Chemical Senses, and Language*. Cornell University Press, Ithaca, NY. The definitive report of von Frisch's experimental analysis of this communication system is von Frisch, K. 1993. *The Dance Language and Orientation of Bees*. Harvard University Press, Cambridge, MA. A recent confirmation of his conclusion that bees share information about rich food sources by means of waggle dances is Riley, J. R., U. Greggers, A. D. Smith, D. R. Reynolds, and R. Menzel. 2005. The flight paths of honeybees recruited by the waggle dance. *Nature* 435:205–207.

Page 9: Quote of Karl von Frisch, from von Frisch, K. 1954. *The Dancing Bees: An Account of the Life and Senses of the Honey Bee*. Methuen, London. Pp. 101, 103.

Page 12: For detailed biographical information on Martin Lindauer, see Seeley, T. D., S. Kühnholz, and R. H. Seeley. 2002. An early chapter in behavioral physiology and sociobiology: the science of Martin Lindauer. *Journal of Comparative Physiology A* 188:439–453.

Page 13: Quote of Martin Lindauer regarding world of humanity, from Seeley, T. D., S. Kühnholz, and R. H. Seeley. 2002. An early chapter in behavioral physiology and sociobiology: the science of Martin Lindauer. *Journal of Comparative Physiology A* 188:439–453. P. 442.

Page 13: Quote of Martin Lindauer regarding beautiful experience, from Seeley, T. D, S. Kühnholz, and R. H. Seeley. 2002. An early chapter in behavioral physiology and sociobiology: the science of Martin Lindauer. *Journal of Comparative Physiology A* 188:439–453. P. 447.

Page 14: Quote of Martin Lindauer regarding dirty dancers, from Lindauer, M. 1955. Schwarmbienen auf Wohnungssuche. *Zeitschrift für vergleichende Physiologie* 37:263–324. P. 266. Translated by T. D. Seeley.

Pages 14–16: Lindauer's discovery that bees can use waggle dances to announce nest sites as well as food sources was first reported in Lindauer, M. 1951. Bienentänze in der Schwarmtraube. *Die Naturwissenschaften* 38:509–513.

Page 17: Roger A. Morse served as professor of apiculture at Cornell University for 40 years, from 1957 to 1997. He supervised the studies of over 30 graduate and postdoctoral students and wrote many leading books on beekeeping, including *The Complete Guide to Beekeeping* (1972, Dutton, New York) and *Bees and Beekeeping* (1975, Cornell University Press, Ithaca, NY).

Page 17: See Wilson, E. O. 1971. *The Insect Societies*. Harvard University Press, Cambridge, MA.

Page 18: See Lindauer, M. 1961. *Communication among Social Bees*. Harvard University Press, Cambridge, MA.

Page 18: Lindauer's magnum opus on house hunting by honeybees is Lindauer, M. 1955. Schwarmbienen auf Wohnungssuche. *Zeitschrift für vergleichende Physiologie* 37:263–324. An English translation, titled House-hunting by honey bee swarms, exists as a supplement to Visscher, P. K. 2007. Group decision making in nest-site selection among social insects. *Annual Review of Entomology* 52:255–275. It is available online at http://arjournals.annualreviews.org/toc/ento/52/1.

Chapter 2. Life in a Honeybee Colony

Page 20: Quote of Charles Butler, from Butler, C. 1609. *The Feminine Monarchie: Or, A Treatise concerning Bees and the Divine Ordering of Them*. Preface, p. 4. Joseph Barnes, Oxford.

Page 20: The most comprehensive "Who's Who" of bees is Michener, C. D. 2000. *The Bees of the World*. Johns Hopkins University Press, Baltimore. For a detailed and beautifully illustrated review of the evolutionary history of bees, see chapter 11, Hymenoptera: ants, bees, and other wasps, in Grimaldi, D., and M. S. Engel. 2005. *Evolution of the Insects*. Cambridge University Press, Cambridge. The recent discovery of the oldest known fossil bee is reported in Poinar, G. O., Jr., and B. N. Danforth. 2006. A fossil bee from Early Cretaceous Burmese amber. *Science* 314:614.

Page 21: The complex mutualism between flowering plants and bees is reviewed in Proctor, M., P. Yeo, and A. Lack. 1996. *The Natural History of Pollination*. Timber Press, Portland, OR. See also Barth, F. G. 1985. *Insects and Flowers: The Biology of a Partnership*. Princeton University Press, Princeton, NJ.

Page 21: The biology of solitary bees is reviewed in comparison to the biology of social bees in Michener, C. D. 1974. *The Social Behavior of the Bees*. Harvard University Press, Cambridge, MA.

Page 21: An observation hive for honeybees is one in which a colony of bees lives between two panes of glass. The hive is built like a sandwich, with the glass for bread and the bees' double-sided comb as filling in the middle. There is a space beneath the glass on each side of the comb so that a single layer of bees can walk around on the comb. Thus all of the hive's inhabitants are always exposed, and a person can peer easily into their normally private world.

Pages 21–25: The anatomy and reproductive biology of workers, queens, and drones are described in detail in Winston, M. L. 1987. *The Biology of the Honey Bee.* Harvard University Press, Cambridge, MA. For an utterly gorgeous description of bee anatomy—with magnificent photographs, micrographs, drawings, and paintings—see Goodman, L. 2003. *Form and Function in the Honey Bee*. International Bee Research Association, Cardiff.

Page 25: The concept of a honeybee colony as a superorganism is developed in Seeley, T. D. 1989. The honey bee colony as a superorganism. *American Scientist* 77:546–553. For a detailed review of the biology of honeybees, with emphasis on how a colony functions as a unified whole, see Moritz, R.F.A., and E. E. Southwick. 1992. *Bees as Superorganisms: An Evolutionary Reality*. Springer-Verlag, Berlin. For a beautiful overview of insect (ant, termite, bee, and wasp) superorganisms, see Hölldobler, B., and E. O. Wilson. 2009. *The Superorganism: The Beauty, Elegance, and Strangeness of Insect Societies.* Norton, New York.

Pages 25–27: For more detailed information on the topics of colony physiology mentioned here, see the following references. Thermoregulation: chapter 16, Social thermoregulation, in Heinrich, B. 1993. *The Hot-Blooded Insects: Strategies and Mechanisms of Thermoregulation.* Harvard University Press, Cambridge, MA. Carbon dioxide regulation: Seeley, T. D. 1974. Atmospheric carbon dioxide regulation in honey bee (*Apis mellifera*) colonies. *Journal of Insect Physiology* 20:2301–2305. Circulation of food: Basile, R., C.W.W. Pirk, and J. Tautz. 2008. Trophallactic activities in the honeybee brood nest—heaters get supplied with high performance fuel. *Zoology* 111:433–441. Fever response: Starks, P. T., C. A. Blackie, and T. D. Seeley. 2000. Fever in honey bee colonies. *Naturwissenschaften* 87:229–231.

Pages 27–33: The annual cycle of honeybee colonies is discussed more thoroughly in chapter 4, The annual cycle of colonies, in Seeley, T. D. 1985. *Honeybee Ecology*. Princeton University Press, Princeton, NJ. See also Seeley, T. D., and P. K. Visscher. 1985. Survival of honeybees in cold climates: the critical timing of colony growth and reproduction. *Ecological Entomology* 10:81–88.

Pages 33–34: The complexities of reproduction by honeybee colonies are described in greater detail in chapter 5, Reproduction, in Seeley, T. D. 1985. *Honeybee Ecology*. Princeton University Press, Princeton, NJ, and in chapter 12, Drones, queens, and mating, in Winston, M. L. 1987. *The Biology of the Honey Bee.* Harvard University Press, Cambridge, MA.

Page 35: The three-year study of the survival, lifespan, reproductive rate, and other demographic characteristics of feral honeybee colonies, is reported in Seeley, T. D. 1978. Life history strategy of the honey bee, *Apis mellifera*. *Oecologia* 32:109–118. An experimental test of the importance of swarming early in the summer is reported in Seeley, T. D., and P. K. Visscher. 1985. Survival of honeybees in cold climates: the critical timing of colony growth and reproduction. *Ecological Entomology* 10:81–88.

Pages 35–42: For a detailed review of the process of honeybee swarming, see chapter 11,

Reproduction: swarming and supersedure, in Winston, M. L. 1987. *The Biology of the Honey Bee*. Harvard University Press, Cambridge, MA.

Page 37: For more information about the curious shaking of the mother queen before her departure in a swarm, and her remarkably effective slimming regime, see Allen, M. D. 1959. The occurrence and possible significance of the "shaking" of honeybee queens by the workers. *Animal Behaviour* 7:66–69; and Pierce, A. L., L. A. Lewis, and S. S. Schneider. 2007. The use of the vibrational signal and worker piping to influence queen behavior during swarming in honey bees, *Apis mellifera*. *Ethology* 113:267–275. Note: the shaking signal is sometimes called the vibration signal or, more awkwardly, the "DVAV," which stands for dorso-ventral abdominal vibration.

Pages 37–38: The first study that documented how worker bees become engorged with honey in preparation for swarming is Combs, G. F., Jr. 1972. The engorgement of swarming worker honeybees. *Journal of Apicultural Research* 11:121–128. More detailed reports on this phenomenon are provided by Otis, G. W., M. L. Winston, and O. R. Taylor, Jr. 1981. Engorgement and dispersal of Africanized honeybee swarms. *Journal of Apicultural Research* 20:3–12; and by Leta, M. A., C. Gilbert, and R. A. Morse. 1996. Levels of hemolymph sugars and body glycogen of honeybees (*Apis mellifera* L.) from colonies preparing to swarm. *Journal of Insect Physiology* 42:239–245. The structure of the wax glands and the physiology of beeswax production are reviewed in chapter 8, Glands: chemical communication and wax production, in Goodman, L. 2003. *Form and Function in the Honey Bee*. International Bee Research Association, Cardiff, and in Hepburn, H. R. 1986. *Honeybees and Wax*. Springer-Verlag, Heidelberg.

Page 38: "The calm before the swarm" is a quote of the worker bee Nyuki, from Hosler, J. 2000. *Clan Apis*. Active Synapse, Columbus, OH. P. 40.

Pages 38–39: For detailed reports on how the scout bees trigger the explosive departure of a swarm from its nest, see Rangel, J., and T. D. Seeley. 2008. The signals initiating the mass exodus of a honeybee swarm from its nest. *Animal Behaviour* 76:1943–1952; Rangel, J., S. R. Griffin, and T. D. Seeley, 2010. An oligarchy of nest-site scouts triggers a honeybee swarm's departure from the hive. *Behavioral Ecology and Sociobiology*, in press. The two signals used by the scout bees—worker piping and buzz running—are described in two papers: Seeley, T. D., and J. Tautz. 2001. Worker piping in honey bee swarms and its role in preparing for liftoff. *Journal of Comparative Physiology A* 187:667–676; and Rittschof, C. C., and T. D. Seeley. 2007. The buzz-run: how honeybees signal "Time to go!" *Animal Behaviour* 75:189–197.

Page 40: The scent organ of worker honeybees, and the chemistry of its attraction pheromones, are reviewed in chapter 13, Attraction: Nasonov pheromone, in Free, J. B. 1987. *Pheromones of Social Bees*. Cornell University Press, Ithaca, NY.

Pages 40–42: For a detailed account of what happens to the virgin queens in a swarming colony after the primary swarm departs, see Gilley, D. C., and D. R. Tarpy. 2005. Three mechanisms of queen elimination in swarming honey bee colonies. *Apidologie* 36:461–474. For a thorough description of the behaviors of queens and workers when the virgin queens are having their fights to the death, see Gilley, D. C. 2001. The behavior of honey bees (*Apis mellifera ligustica*) during queen duels. *Ethology* 107:601–622. An analysis of the adaptive design of the fighting behavior of virgin queens is provided in Visscher, P. K. 1993. A theoretical analysis of individual interests and intracolony conflict during swarming of honey bee colonies. *Journal of Theoretical Biology* 165:191–212.

Page 41: The toots and quacks of queen honeybees are described precisely, based on laser vibrometer recordings, in Michelsen, A., W. H. Kirchner, B. B. Andersen, and M. Lindauer. 1986. The tooting and quacking vibration signals of honeybee queens: a quantitative analysis. *Journal of Comparative Physiology A* 158:605–611. For a general review of the diverse acoustical signals, both sounds and vibrations, that honeybees use in communicating in the darkness inside a hive, see Kirchner, W. H. 1993. Acoustical communication in honeybees. *Apidologie* 24:297–307.

Chapter 3. Dream Home for Honeybees

Page 43: Quote of Robert Frost, from "A Drumlin Woodchuck," in Latham, E. C., ed. 1969. *The Poetry of Robert Frost*. Henry Holt, New York.

Pages 43–44: The history of mankind's association with bees, as evidenced by material objects (cave paintings, illuminated manuscripts, hives and their shelters, and beekeeping tools), is explored in Crane, E. 1983. *The Archaeology of Beekeeping*. Duckworth, London.

Page 45: The first experimental studies of the nest-site preferences of honeybees are reported in Lindauer, M. 1955. Schwarmbienen auf Wohnungssuche. *Zeitschrift für vergleichende Physiologie* 37:263–324. An English translation, titled House-hunting by honey bee swarms, exists as a supplement to Visscher, P. K. 2007. Group decision making in nest-site selection among social insects. *Annual Review of Entomology* 52:255–275. It is available online at http://arjournals.annualreviews.org/toc/ento/52/1.

Page 45: Quote of Martin Lindauer about asking the bees themselves, from Lindauer, M. 1955. Schwarmbienen auf Wohnungssuche. *Zeitschrift für vergleichende Physiologie* 37:263–324. P. 290. Translated by P. K. Visscher.

Page 47: Quote about the von Frisch–Lindauer approach to animal behavior research from Hölldobler, B., and E. O. Wilson. 1994. *Journey to the Ants*. Harvard University Press, Cambridge, MA. P. 19.

Pages 49–51: For a detailed report on the nests of honeybees living in trees, see Seeley, T. D., and R. A. Morse. 1976. The nest of the honey bee (*Apis mellifera* L.). *Insectes Sociaux* 23:495–512.

Page 51: Forest beekeeping in medieval Russia, Poland, Germany, and England is reviewed in chapter 5, Forest "beekeeping" and the precursor of upright hives, in Crane, E. 1983. *The Archaeology of Beekeeping*. Duckworth, London. See also Galton, D. 1971. *Survey of a Thousand Years of Beekeeping in Russia*. Bee Research Association, London.

Page 52: The craft of lining bees is best described in Edgell, G. H. 1949. *The Bee Hunter*. Harvard University Press, Cambridge, MA. Edgell, who was director of the Museum of Fine Arts in Boston, had hunted bee trees in New Hampshire since his boyhood and once said that his book on bee hunting brought him greater fame than his professional publications on fine arts. I have used his no-nonsense methods for hunting wild colonies of bees in several ecological studies. See Visscher, P. K., and T. D. Seeley. 1982. Foraging strategy of honeybee colonies in a temperate deciduous forest. *Ecology* 63:1790–1801; Seeley, T. D. 2007. Honey bees of the Arnot Forest: a population of feral colonies persisting with *Varroa destructor* in the northeastern United States. *Apidologie* 38:19–29; and Seeley, T. D. 2008. The bees of the Arnot Forest. *Bee Culture* 136 (March):23–25.

Page 53: The use of bait hives in African beekeeping is described in Smith, F. G. 1960. *Beekeeping in the Tropics*. Longmans, London; and Guy, R. D. 1972. Commercial beekeeping with African bees. *Bee World* 53:14–22.

Pages 54–58: Full reports of my studies of the bees' nest-site preferences are found in Seeley, T. D. 1977. Measurement of nest cavity volume by the honey bee (*Apis mellifera*). *Behavioral Ecology and Sociobiology* 2:201–227; Seeley, T. D., and R. A. Morse. 1978. Nest site selection by the honey bee *Apis mellifera*. *Insectes Sociaux* 25:323–337; and Visscher, P. K., R. A. Morse, and T. D. Seeley. 1985. Honey bees choosing a home prefer previously occupied cavities. *Insectes Sociaux* 32:217–220. Other papers on the same topic include Jaycox, E. R., and S. G. Parise. 1980. Homesite selection by Italian honey bee swarms, *Apis mellifera ligustica* (Hymenoptera: Apidae). *Journal of the Kansas Entomological Society* 53:171–178; Jaycox, E. R., and S. G. Parise. 1981. Homesite selection by swarms of black-bodied honey bees, *Apis mellifera caucasica* and *A. m. carnica* (Hymenoptera: Apidae). *Journal of the Kansas Entomological Society* 54:697–703; and Rinderer, T. E., K. W. Tucker, and A. M. Collins. 1982. Nest cavity selection by swarms of European and Africanized honeybees. *Journal of Apicultural Research* 21:98–103.

Page 55: The effects of entrance direction on the overwintering success of colonies is reported in Szabo, T. I. 1983. Effects of various entrances and hive direction on outdoor wintering of honey bee colonies. *American Bee Journal* 123:47–49.

Page 56: The study of the size distribution of natural tree cavities in a Vermont forest is reported in Seeley, T. D. 1977. Measurement of nest cavity volume by the honey bee (*Apis mellifera*). *Behavioral Ecology and Sociobiology* 2:201–227.

Page 57: The economics of comb construction are reviewed in chapter 6, Nest building, in Seeley, T. D. 1985. *Honeybee Ecology*. Princeton University Press, Princeton, NJ.

Page 57: The extensive use of tree resins ("propolis") by honeybees to seal their nesting cavities is described in Seeley, T. D., and R. A. Morse. 1976. The nest of the honey bee (*Apis mellifera* L.). *Insectes Sociaux* 23:495–512. How the bees handle resins within the nest and how a colony controls its resin collection is reported in Nakamura, J., and T. D. Seeley. 2006. The functional organization of resin work in honeybee colonies. *Behavioral Ecology and Sociobiology* 60:339–349.

Pages 58–59: The study of the Asian honeybees in Thailand is reported in Seeley, T. D., R. H. Seeley, and P. Akratanakul. 1982. Colony defense strategies of the honeybees in Thailand. *Ecological Monographs* 52:43–63.

Page 59: For detailed information on the Yellow Rain story, see Seeley, T. D., J. Nowicke, M. Meselson, J. Guillemin, and P. Akratanakul. 1985. Yellow rain. *Scientific American* 253 (September):128–137.

Page 59: KGB is the Russian acronym for the Committee of State Security (i.e., the national security agency) of the USSR. From 1954 to 1991, it was the communist state's premier secret police and intelligence organization.

Page 60: For information on bait hives, see Morse, R. A. and T. D. Seeley. 1978. Bait hives. *Gleanings in Bee Culture* 106 (May):218–220, 242; Morse, R. A., and T. D. Seeley. 1979. New observations on bait hives. *Gleanings in Bee Culture* 107 (June):310–311, 327; Seeley, T. D., and R. A. Morse. 1982. Bait hives for honey bees. *Cornell Cooperative Extension Information Bulletin* No. 107; Witherell, P. C. 1985. A review of the scientific literature relating to honey bee bait hives and swarm attractants. *American Bee Journal* 125:823–829; Ratnieks, F.L.W. 1988. Improved bait hives. *American Bee Journal* 128:125–127; and Schmidt, J. O., S. C. Thoenes, and R. Hurley. 1989. Swarm traps. *American Bee Journal* 129:468–471.

Page 60: For information on attracting swarms to bait hives using attraction pheromones, see Free, J. B., J. A. Pickett, A. W. Ferguson, and M. C. Smith. 1981. Synthetic pheromones to attract honeybee (*Apis mellifera*) swarms. *Journal of Agricultural Science* 97:427–431; Schmidt, J. O., K. N. Slessor, and M. L. Winston. 1993. Roles of Nasonov and queen pheromones in attraction of honeybee swarms. *Naturwissenschaften* 80:573–575; Winston, M. L., K. N. Slessor, W. L. Rubink, and J. D. Villa. 1993. Enhancing pheromone lures to attract honey bee swarms. *American Bee Journal* 133:58–60; and Schmidt, J. O. 1994. Attraction of reproductive honey bee swarms to artificial nests by Nasonov pheromone. *Journal of Chemical Ecology* 20:1053–1056.

Page 63: The character of the Isles of Shoals (Gulf of Maine) and the construction of the Shoals Marine Laboratory on Appledore Island in the 1960s and 1970s are described in Kingsbury, J. M. 1991. *Here's How We'll Do It*. Bullbrier Press, Ithaca, NY.

Pages 67–71: For a detailed report on the nest-site inspection behavior of scout bees and on the experimental analysis of how a small bee measures the volume of a large space, see

Seeley, T. D. 1977. Measurement of nest cavity volume by the honey bee (*Apis mellifera*). *Behavioral Ecology and Sociobiology* 2:201–227.

Pages 71–72: The nifty algorithm, suggested by Nigel Franks and Anna Dornhaus, by which bees might measure the volumes of potential nesting cavities is described in Franks, N. R., and A. Dornhaus. 2003. How might individual honeybees measure massive volumes? *Proceedings of the Royal Society of London B (Supplement)* 270, S181–S182.

Chapter 4. Scout Bees' Debate

Page 73: Quote of Jimmy Carter, from Carter, J. E. 1978. *Address to the Parliament of India*, June 2, 1978.

Pages 73–74: The New England town meeting is a fascinating form of small town democratic government. How it works is described in Mansbridge, J. J. 1983. *Beyond Adversary Democracy*. University of Chicago Press, Chicago; and Bryan, F. M. 2004. *Real Democracy.* University of Chicago Press, Chicago.

Pages 75–84: The full report of Lindauer's observations of dancing bees on swarm clusters is found in pages 265–282 in his magnum opus: Lindauer, M. 1955. Schwarmbienen auf Wohnungssuche. *Zeitschrift für vergleichende Physiologie* 37:263–324. An English translation, titled House-hunting by honey bee swarms, exists as a supplement to Visscher, P. K. 2007. Group decision making in nest-site selection among social insects. *Annual Review of Entomology* 52:255–275. It is available online at http://arjournals.annualreviews.org/toc/ento/52/1.

Page 75: The phrase "watching and wondering" comes from the title of an autobiography written by Niko Tinbergen. See Tinbergen, N. 1985. Watching and wondering, in Dewsbury, D. A., ed. *Studying Animal Behavior: Autobiographies of the Founders*. University of Chicago Press, Chicago. Pp. 431–463. Tinbergen strongly advocated starting a study of animal behavior by conducting a descriptive reconnaissance of the behavior to get a broad view of the phenomenon.

Page 76: The clever system devised by Karl von Frisch for making hundreds of bees individually identifiable using paint dots of just five colors is described in von Frisch, K. 1993. *The Dance Language and Orientation of Bees*. Harvard University Press, Cambridge, MA. Pp. 14–17.

Page 81: Quote of Martin Lindauer regarding a tug-of-war between two groups of dancers, from Lindauer, M. 1955. Schwarmbienen auf Wohnungssuche. *Zeitschrift für vergleichende Physiologie* 37:263–324. P. 276. Translated by P. K. Visscher.

Page 82: Lindauer describes two instances of a flying swarm making an emergency stopover while en route to its new home. See Lindauer, M. 1955. Schwarmbienen auf Wohnungssuche. *Zeitschrift für vergleichende Physiologie* 37:263–324. Pp. 319–320.

Page 82: Quote of Martin Lindauer regarding a swarm dividing itself, from Lindauer, M. 1961. *Communication among Social Bees.* Harvard University Press, Cambridge, MA. P. 45.

Page 85: Quote of Martin Lindauer regarding scouts affiliated with losing sites giving up recruitment, from Lindauer, M. 1955. Schwarmbienen auf Wohnungssuche. *Zeitschrift für vergleichende Physiologie* 37:263–324. P. 275. Translated by P. K. Visscher.

Page 86: For a good discussion of the distinction between the two categories of group choice—consensus vs. combined—see Conradt, L., and T. J. Roper. 2005. Consensus decision making in animals. *Trends in Ecology and Evolution* 20:449–456; and Conradt, L., and C. List. 2009. Introduction. Group decisions in humans and animals: a survey. *Philosophical Transactions of the Royal Society B* 364:719–742.

Page 86: The full reference to the book mentioned here is Seeley, T. D. 1995. *The Wisdom of the Hive: The Social Physiology of Honey Bee Colonies*. Harvard University Press, Cambridge, MA.

Pages 86–92: For the full report of the eavesdropping by Susannah Buhrman and me on the scout bees' debates on three swarms, see Seeley, T. D., and S. C. Buhrman. 1999. Group decision making in swarms of honey bees. *Behavioral Ecology and Sociobiology* 45:19–31.

Pages 93–94: Lindauer's studies of which bees take up the profession of nest-site scout and when they do so are reported in Lindauer, M. 1955. Schwarmbienen auf Wohnungssuche. *Zeitschrift für vergleichende Physiologie* 37:263–324. Pp. 296–307.

Page 94: Quote of Martin Lindauer regarding irresolute foragers, from Lindauer, M. 1955. Schwarmbienen auf Wohnungssuche. *Zeitschrift für vergleichende Physiologie* 37:263–324. P. 306. Translated by P. K. Visscher.

Pages 95–96: For a detailed report on Dave Gilley's study of the striking age distribution of the nest-site scouts, see Gilley, D. C. 1998. The identity of nest-site scouts in honey bee swarms. *Apidologie* 29:229–240.

Page 96: For up-to-date discussions of how nature and nurture interact in shaping the complex social behavior of honeybees, see Robinson, G. E. 2004. Beyond nature and nurture. *Science* 304:397–399; and Robinson, G. E. 2006. Genes and social behaviour, in Lucas, J. R., and L. W. Simmons, eds. *Essays in Animal Behaviour: Celebrating 50 Years of Animal Behaviour*. Elsevier, London. Pp. 101–113.

Pages 96–97: For the full report of Gene Robinson's and Robert Page's test for a genetic influence on a bee's likelihood of becoming a nest-site scout, see Robinson, G. E., and R. E. Page, Jr. 1989. Genetic determination of nectar foraging, pollen foraging, and nest-site scouting in honey bee colonies. *Behavioral Ecology and Sociobiology* 24:317–323.

Page 97: For a more detailed protocol for preparing artificial swarms, see the methods section in Seeley, T. D. 2003. Consensus building during nest-site selection in honey bee swarms: the expiration of dissent. *Behavioral Ecology and Sociobiology* 53:417–424.

Page 98: Lindauer's observations on what foragers were experiencing when they started scouting are reported in Lindauer, M. 1955. Schwarmbienen auf Wohnungssuche. *Zeitschrift für vergleichende Physiologie* 37:263–324. Pp. 304–306.

Page 98: For detailed reports on how nectar foragers lose their enthusiasm for dancing and foraging when they have difficulty finding hive bees to take their nectar loads, see Seeley, T. D. 1989. Social foraging in honey bees: how nectar foragers assess their colony's nutritional status. *Behavioral Ecology and Sociobiology* 24:181–199; and Seeley, T. D., and C. A. Tovey. 1994. Why search time to find a food-storer bee accurately indicates the relative rates of nectar collecting and nectar processing in honey bee colonies. *Animal Behaviour* 47:311–316.

Chapter 5. Agreement on Best Site

Page 99: Quote of John Milton, from Milton, J. 1671. *Samson Agonistes*. Line 1008.

Page 100: The economist Herbert A. Simon proposed the concept of bounded rationality in the mid 1950s in Simon, H. A. 1956. Rational choice and structure of environments. *Psychological Review* 63:129–138; and in Simon, H. A. *Models of Man*. Wiley, New York. For a recent book on the topic that includes several chapters on decision-making heuristics, see Gigerenzer, G., and R. Selten. 2001. *Bounded Rationality: The Adaptive Toolbox*. MIT Press, Cambridge, MA.

Page 100: A good discussion of one-reason decision making is provided by Gigerenzer, G., and D. G. Goldstein. 1999. Betting on one good reason: take the best and its relatives, in Gigerenzer, G., P. M. Todd, and The ABC Research Group, eds. *Simple Heuristics That Make Us Smart*. Oxford University Press, New York. Pp. 75–95.

Page 102: I believe that Susannah Buhrman and I observed a case of a first-rate site getting entered so late in a scout bees' debate that the swarm rejected this excellent site in favor of a poorer one. This occurred when we watched the debate depicted in figure 4.6. Well into this debate, at 2:49 p.m. on June 20, the scout bee Green-White 39 landed on the swarm and in great excitement performed a lively and lengthy dance (166 dance circuits) for site L, just 200 meters (650 feet) to the southwest. She had discovered in a large white pine tree the empty nest of a wild colony that had died out over the preceding winter . . . a superb home! At 3:50 she again landed on the swarm and excitedly performed a second long-lasting dance (95 dance circuits). Meanwhile, however, dozens of other scouts were dancing rather faintheartedly for site I, 4,200 meters (13,800 feet, or 2.6 miles) to the south. Their dances contained on average only 6 dance circuits, indicating that site I was much less desirable than site L. Nevertheless, the swarm chose site I rather than site L, evidently because the excellent news brought back by Green-White 39 was too little and too late to redirect the swarm's debate in favor of her superior alternative.

Page 102: The "secrets about the beautiful inner workings of a honeybee colony" that were uncovered at the Cranberry Lake Biological Station are described in Seeley, T. D. 1995. *The Wisdom of the Hive*. Harvard University Press, Cambridge, MA. This is a book worth reading!

Page 103–110: For the full report on the experiments that were conducted to determine how to create a mediocre, but acceptable, artificial nest site, see Seeley, T. D., and S. C. Buhrman. 2001. Nest-site selection in honey bees: how well do swarms implement the "Best-of-N" decision rule? *Behavioral Ecology and Sociobiology* 49:416–427.

Page 107: The specific procedure that we used for counting the number of scouts at a nest box was as follows. The person making the count sat 3 meters (about 10 feet) in front of the box and over a three-minute period made five counts of the maximum number of bees seen simultaneously at the box (either flying around it or crawling upon it) during a 30-second period. We used the average of these five counts as our measure of the scout bees' interest in the box at that time.

Page 111: The evidence that the strength of a honeybee's waggle dance provides an accurate readout of her evaluation of the desirability of a food source comes from several studies, including Waddington, K. D. 1982. Honey bee foraging profitability and round dance correlates. *Journal of Comparative Physiology* 148:297–301; Seeley, T. D. 1994. Honey bee foragers as sensory units of their colonies. *Behavioral Ecology and Sociobiology* 34:51–62; and Seeley, T. D., A. S. Mikheyev, and G. J. Pagano. 2000. Dancing bees tune both duration and rate of waggle-run production in relation to nectar-source profitability. *Journal of Comparative Physiology A* 186:813–819.

Pages 111–115: For the detailed description of the best-of-5 choice test, see Seeley, T. D., and S. C. Buhrman. 2001. Nest-site selection in honey bees: how well do swarms implement the "Best-of-N" decision rule? *Behavioral Ecology and Sociobiology* 49:416–427.

Page 116–117: My study of honeybee colony survival as a function of nest cavity volume, which tests whether the bees' nest-site preferences are beneficial to them, is not yet published. Some other studies that have likewise examined whether the nest-site preferences of animals enhance their reproductive success include Courtenay, S. C., and M.H.A. Keenleyside. 1983. Nest site selection by the fourspine stickleback, *Apeltes quadracus* (Mitchell). *Canadian Journal of Zoology* 61:1443–1447; Morse, D. H. 1985. Nests and nest-site selection of the crab spider *Misumena vatia* (Araneae, Thomisidae) on milkweed. *Journal of Arachnology* 13:383–390; Regehr, H. M., M. S. Rodway, and W. A. Montevecchi. 1998. Antipredator benefits of nest-site selection in black-legged kittiwakes. *Canadian Journal of Zoology* 76:910–913; and Wilson, D. S. 1998. Nest-site selection: microhabitat variation and its effects on the survival of turtle embryos. *Ecology* 79:1884–1892.

Chapter 6. Building a Consensus

Page 118: Quote from Society of Friends. 1934. *Book of Discipline*. Part I. Friends' Book Centre, London.

Page 118–119: The recognition that democracy can come in two different forms—

adversary and unitary—and that the mechanisms of decision making differ markedly between them, was first made in Mansbridge, J. J. 1983. *Beyond Adversary Democracy.* University of Chicago Press, Chicago. Mansbridge points out that the unitary process of decision making, unlike the adversary one, consists not in counting votes made by secret ballot to identify the majority opinion, but in an open and direct discussion to build a consensus. Clearly, the honeybee's house-hunting process is an example of unitary democracy.

Pages 120–121: The analogy between the decision-making process of honeybee swarms and the democratic election process of our own societies was originally drawn in Britton, N. F., N. R. Franks, S. C. Pratt, and T. D. Seeley. 2002. Deciding on a new home: how do honeybees agree? *Proceedings of the Royal Society of London B* 269:1383–1388. The main aim of this paper was, however, to extend classical mathematical models of the spread of infectious diseases, and of infectious ideas, to elucidate the decision-making process of house-hunting honeybees. This theoretical work shows that there is no need for any bee to make comparisons between sites. Later empirical work (described later in this chapter) showed that indeed scout bees do not make comparisons of sites.

Pages 121–122: Lindauer's observations on how a nest-site scout adjusts the strength of her dance according to the quality of the site she is advertising are reported in Lindauer, M. 1955. Schwarmbienen auf Wohnungssuche. *Zeitschrift für vergleichende Physiologie* 37:263–324. Pp. 294–296.

Page 122: Quote of Martin Lindauer regarding lively versus lackluster dances, from Lindauer, M. 1955. Schwarmbienen auf Wohnungssuche. *Zeitschrift für vergleichende Physiologie* 37:263–324. P. 296. Translated by P. K. Visscher.

Page 123: The first quantitative evidence that scouts advertise superior nest sites with longer and livelier dances is reported in Seeley, T. D., and S. C. Buhrman. 2001. Nest-site selection in honey bees: how well do swarms implement the "Best-of-N" decision rule? *Behavioral Ecology and Sociobiology* 49:416–427. The parallel finding that nectar foragers advertise richer flower patches by increasing the duration (= dance length) and rate (= dance liveliness) of dance-circuit production is reported in Seeley, T. D., A. S. Mikheyev, and G. J. Pagano. 2000. Dancing bees tune both duration and rate of waggle-run production in relation to nectar-source profitability. *Journal of Comparative Physiology A* 186:813–819.

Page 123: Quote of Martin Lindauer regarding faint-hearted dance etc., from Lindauer, M. 1961. *Communication among Social Bees.* Harvard University Press, Cambridge, MA. P. 49.

Page 125–126: Kirk Visscher perfected the "abduction by aliens" method of labeling scout bees upon exiting a nest box when he performed some of the studies reported in Visscher, P. K. and S. Camazine. 1999. Collective decisions and cognition in bees. *Nature* 397:400.

Page 126–128: For the full report on how scouts behave differently when reporting on a high-quality (40-liter) nesting site versus a medium-quality (15-liter) one, see Seeley, T. D., and P. K. Visscher. 2008. Sensory coding of nest-site value in honeybee swarms. *Journal of Experimental Biology* 211:3691–3697.

Page 129: A more complete analysis of how the problem of noisy individual-level coding of site quality decreases as the number of scouts reporting on a site increases is found in Seeley, T. D., and P. K. Visscher. 2008. Sensory coding of nest-site value in honeybee swarms. *Journal of Experimental Biology* 211:3691–3697.

Page 129: The handy expression "force of persuasion," which is the product of the number of bees dancing for a site and the average number of dance circuits produced per dancing bee, comes from Britton, N. F., N. R. Franks, S. C. Pratt, and T. D. Seeley. 2002. Deciding on a new home: how do honeybees agree? *Proceedings of the Royal Society of London B* 269:1383–1388. It is analogous to the epidemiological "force of infection" in mathematical models for the spread of infectious diseases.

Page 130: The evidence regarding the nest-site scouts' rule that a "discoverer-should-dance" is reported in Seeley, T. D., and P. K. Visscher. 2008. Sensory coding of nest-site value in honeybee swarms. *Journal of Experimental Biology* 211:3691–3697.

Page 131: Several studies have carefully examined whether worker honeybees are genetically endowed with innate preferences, especially regarding the stimuli representing flowers. For reviews of classic studies on the innate search images guiding the first foraging flights of bees, see Menzel, R. 1985. Learning in honey bees in an ecological and behavioral context, in Hölldobler, B., and M. Lindauer, eds. *Experimental Behavioral Ecology and Sociobiology*. Gustav Fischer Verlag, Stuttgart. Pp. 55–74; and Gould, J. L., and W. F. Towne. 1987. Honey bee learning. *Advances in Insect Physiology* 20:55–75. For more recent original studies, see Giurfa, M., J. A. Núñez, L. Chittka, and R. Menzel. 1995. Colour preferences of flower-naive honeybees. *Journal of Comparative Physiology A* 177:247–259; Rodriguez, I., A. Gumbert, N. Hempel de Ibarra, J. Kunze, and M. Giurfa. 2004. Symmetry is in the eye of the "beholder": innate preference for bilateral symmetry in flower-naive bumblebees. *Naturwissenschaften* 91:374–377.

Pages 134–135: For the full report of the study that found that neutral scouts evidently follow dances at random as they get converted into supporters, see Visscher, P. K., and S. Camazine. 1999. Collective decisions and cognition in bees. *Nature* 397:400. See also Camazine, S., P. K. Visscher, J. Finley, and R. S. Vetter. 1999. House-hunting by honey bee swarms: collective decision and individual behaviors. *Insectes Sociaux* 46:348–360.

Page 135: Mary R. Myerscough, a mathematical biologist at the University of Sydney, Australia, has created a Leslie matrix model of the population dynamics of scout bees performing dances for different nest sites. She has proven, quite elegantly, that given enough time, the dancing scouts in a swarm will almost always become focused on the one best site that has been found. See Myerscough, M. R. 2003. Dancing for a decision: a matrix model for nest-site choice by honey bees. *Proceedings of the Royal Society of London B* 270:577–582.

Page 136: Quote of Martin Lindauer regarding how scout bees lose interest in a lesser nest site, from Lindauer, M. 1955. Schwarmbienen auf Wohnungssuche. *Zeitschrift für vergleichende Physiologie* 37:263–324. P. 296. Translated by P. K. Visscher. Lindauer repeated in later reports his view that scouts cease advertising poorer sites after being recruited to a new and superior site and comparing the new and old sites. For example, in 1957 he wrote, "When, furthermore, those scout bees which at first had announced the inferior nesting places are won over by the more lively dances of their competitors and as a result themselves inspect this home—so that they can compare the two—then they naturally choose the better one. Hence nothing more stands in the way of an agreement." See Lindauer, M. 1957. Communication in swarm-bees searching for a new home. *Nature* 179:63–66. P. 64.

Page 136: Quote of Martin Lindauer regarding nest-site scouts not being stubborn, from Lindauer, M. 1955. Schwarmbienen auf Wohnungssuche. *Zeitschrift für vergleichende Physiologie* 37:263–324. P. 312. Translated by P. K. Visscher. Quote of Martin Lindauer about scouts letting their minds be changed, from Lindauer, M. 1961. *Communication among Social Bees.* Harvard University Press, Cambridge, MA. P. 49.

Page 137: Quote of Martin Lindauer regarding scout bees that ceased to dance for one site before they had inspected another housing possibility, from Lindauer, M. 1955. Schwarmbienen auf Wohnungssuche. *Zeitschrift für vergleichende Physiologie* 37:263–324. P. 296. Translated by P. K. Visscher.

Page 137: For a detailed discussion of the power of figuring things out by testing the predictions of competing hypotheses, see Platt, J. R. 1964. Strong inference. *Science* 146:347–353.

Pages 137–144: For the detailed report of the test between the compare-and-convert vs. the retire-and-rest hypotheses, see Seeley, T. D. 2003. Consensus building during nest-site selection in honey bee swarms: the expiration of dissent. *Behavioral Ecology and Sociobiology* 53:417–424.

Page 145: Quote of Max Planck on the importance of turnover of scientists for the development of science, from Planck, M. 1950. *Scientific Autobiography and Other Papers*. Translated by F. Gaynor. Williams and Norgate, London. P. 33.

Page 145: For a thorough discussion of how new theories get accepted within a scientific community (i.e., how scientists make group decisions about new ideas), see Hull, D. L. 1988. *Science as a Process*. University of Chicago Press, Chicago.

Chapter 7. Initiating the Move to New Home

Page 146: Quote of Charles Butler, from Butler, C. 1609. *The Feminine Monarchie: Or, A Treatise concerning Bees and the Divine Ordering of Them.* Chapter 5, p. 14. Joseph Barnes, Oxford.

Pages 148–151: For the detailed report on Bernd Heinrich's study of temperature regulation in honeybee swarms, see Heinrich, B. 1981. The mechanisms and energetics of honeybee swarm temperature regulation. *Journal of Experimental Biology* 91:25–55. He has also written a broad review of thermoregulation in insects in general. See Heinrich, B. 1993. *The Hot-Blooded Insects*. Harvard University Press, Cambridge, MA.

Page 150: For a detailed look at how the mantle (outermost) workers in a bivouacked swarm of honeybees adjust their body orientation, wing spread, and interindividual spacing to reduce convective heat loss from the swarm cluster, see Cully, S. M., and T. D. Seeley. 2004. Self-assemblage formation in a social insect: the protective curtain of a honey bee swarm. *Insectes Sociaux* 51:317–324.

Pages 152–154: For the detailed report on the thermographic study of swarms warming up in preparation for flight, see Seeley, T. D., M. Kleinhenz, B. Bujok, and J. Tautz. 2003. Thorough warm-up before take-off in honey bee swarms. *Naturwissenschaften* 90:256–260.

Page 155: Quote of Martin Lindauer regarding which bees in a swarm produce the high-pitched piping sounds, from Lindauer, M. 1955. Schwarmbienen auf Wohnungssuche. *Zeitschrift für vergleichende Physiologie* 37:263–324. P. 317. Translated by P. K. Visscher.

Pages 156–162: For the detailed report of the study of scout bees producing the piping signal, see Seeley, T. D., and J. Tautz. 2001. Worker piping in honey bee swarms and its role in preparing for liftoff. *Journal of Comparative Physiology A* 187:667–676.

Page 160: For more information on the role of the shaking or vibration signal in honeybee swarms, see Schneider, S. S., P. K. Visscher, and S. Camazine. 1998. Vibration signal behavior of waggle-dancers in swarms of the honey bee, *Apis mellifera* (Hymenoptera: Apidae). *Ethology* 104:963–972; Lewis, L. A., and S. S. Schneider. 2000. The modulation of worker behavior by the vibration signal during house hunting in swarms of the honeybee, *Apis mellifera*. *Behavioral Ecology and Sociobiology* 48:154–164; Donahoe, K., L. A. Lewis, and S. S. Schneider. 2003. The role of the vibration signal in the house-hunting process of honey bee (*Apis mellifera*) swarms. *Behavioral Ecology and Sociobiology* 54:593–600; and Pierce, A. L., L. A. Lewis, and S. S. Schneider. 2007. The use of the vibration signal and worker piping to influence queen behavior during swarming in honey bees, *Apis mellifera*. *Ethology* 113:267–275.

Pages 163–165: For the detailed report on the form and function of the buzz-run, see Rittschof, C. C., and T. D. Seeley. 2008. The buzz-run: how honeybees signal "Time to go!" *Animal Behaviour* 75:189–197.

Page 165: The classic paper on the origins and evolution of communication signals, through the process of ritualization, is Tinbergen, N. 1952. "Derived" activities: their causation, biological significance, origin, and emancipation during evolution. *Quarterly Review of Biology* 27:1–32. An up-to-date treatment of signal evolution is Bradbury, J. W., and S. L. Vehrencamp. 1998. *Principles of Animal Communication*. Sinauer, Sunderland, MA.

Pages 166–167: Another possible example of a control system in a large social insect colony that is based on a small subset of individuals collecting information about colony state and then, when the time is right, signaling when to take action has been reported for the display tournaments between competing colonies of the honeypot ant, *Myrmecocystus mimicus*. See Lumsden, C. J., and B. Hölldobler. 1983. Ritualized combat and intercolony communication in ants. *Journal of Theoretical Biology* 100:81–98.

Pages 168–172: For detailed reports on the collaborative studies conducted by Kirk Visscher and me, to test for consensus sensing or quorum sensing, see Seeley, T. D., and P. K. Visscher. 2003. Choosing a home: how the scouts in a honey bee swarm perceive the completion of their group decision making. *Behavioral Ecology and Sociobiology* 54:511–520; and Seeley, T. D., and P. K. Visscher. 2004. Quorum sensing during nest-site selection by honeybee swarms. *Behavioral Ecology and Sociobiology* 56:594–601.

Page 171: For the detailed report on which bees produce the piping signal (only scout bees and only ones from the site where a quorum of scouts has formed), see Visscher, P. K., and T. D. Seeley. 2007. Coordinating a group departure: who produces the piping signals on honeybee swarms? *Behavioral Ecology and Sociobiology* 61:1615–1621.

Pages 173–174: Decision makers often face the problem of finding a suitable compromise between speedy decisions and accurate ones. The trade-off between speed and accuracy arises because if an animal has to make a fast decision then it may be prone to make a poor decision, either because it cannot sample its options sufficiently broadly or because it cannot deliberate on its options sufficiently deeply. For a recent review of this topic, see Chittka, L., P. Skorupski, and N. E. Raine. 2009. Speed-accuracy trade-offs in animal decision making. *Trends in Ecology and Evolution* 24:400–407. For a specific study with humans, see Osman, A., L. G. Lou, H. Muller-Gethman, G. Rinkenauer, S. Mattes, and R. Ulrich. 2000. Mechanisms of speed-accuracy trade-off: evidence from covert motor processes. *Biological Psychology* 51:173–199. For one with bees, see Chittka, L., A. G. Dyer, F. Bock, and A. Dornhaus. 2003. Bees trade-off foraging speed for accuracy. *Nature* 424:388. This trade-off has also been demonstrated in ant colonies. See Franks, N. R., A. Dornhaus, J. P. Fitzsimmons, and M. Stevens. 2003. Speed versus accuracy in collective decision making. *Proceedings of the Royal Society of London B* 270:2457–2463.

Page 174: The Quaker method of making group decisions by consensus is described in Pollard, F. E., B. E. Pollard, and R.S.W. Pollard. 1949. *Democracy and the Quaker Method*. Bannisdale Press, London.

Chapter 8. Steering the Flying Swarm

Page 175: Quote of Thomas Smibert, from "The Wild Earth-Bee," in Smibert, T. 1851. *Io Anche! Poems, Chiefly Lyrical*. James Hogg, Edinburgh.

Page 175: The mechanisms used by honeybees to navigate to distant flowers and then find their way home are reviewed in Collett, T. S., and M. Collett. 2002. Memory use in insect visual navigation. *Nature Reviews Neuroscience* 3:542–552; Dyer, F. C. 1998. Spatial cognition: lessons from central-place foraging insects, in Balda, R. P., I. M. Pepperberg, and A. C. Kamil, eds. *Animal Cognition in Nature*. Academic Press, New York, Pp. 119–154; Menzel, R., and M. Giurfa. 2006. Dimensions of cognition in an insect, the honeybee. *Behavioral and Cognitive Neuroscience Reviews* 5:24–40; and Wehner, R. 1992. Arthropods, in Papi, F., ed. *Animal Homing*. Chapman and Hall, London. Pp. 45–144.

Pages 176–177: For the detailed report of the study that demonstrated that worker bees in a flying swarm sense the presence of the queen by smelling the 9-ODA that she produces, see Avitabile, A., R. A. Morse, and R. Boch. 1975. Swarming honey bees guided by pheromones. *Annals of the Entomological Society of America* 68:1079–1082.

Pages 177–179: The full description of the flight of the swarm across Appledore Island in 1979 is found in Seeley, T. D., R. A. Morse, and P. K. Visscher. 1979. The natural history of the flight of honey bee swarms. *Psyche* 86:103–113.

Pages 179–182: For the detailed report on the behaviors of the flying swarms that were tracked for 270 meters at the Liddell Field Station, see Beekman, M., R. L. Fathke, and T. D. Seeley. 2006. How does an informed minority of scouts guide a honey bee swarm as it flies to its new home? *Animal Behaviour* 71:161–171. This paper also reports the high flight speeds of other swarms that were tracked as they made flights of 1,000 and 4,000 meters.

Page 182: Honeybee swarms are not unique in being guided toward a target by a small fraction of the group's members who are informed about the target's location. For experimental studies demonstrating this in fish schools and human groups, see Reebs, S. G. 2000. Can a minority of informed leaders determine the foraging movements of a fish shoal? *Animal Behaviour* 59:403–409; Ward, A.J.W., D.J.T. Sumpter, I. D. Couzin, P.J.B. Hart, and J. Krause. 2008. Quorum decision making facilitates information transfer in fish shoals. *Proceedings of the National Academy of Sciences, U.S.A.* 105:6948–6953; and Dyer, J.R.G., C. C. Ioannou, L. J. Morrell, D. P. Croft, I. D. Couzin, D. A. Waters, and J. Krause. 2008. Consensus decision making in human crowds. *Animal Behaviour* 75:461–470.

Pages 182–183: The hypothesis that scouts guide the flight of a swarm with pheromones was proposed in Avitabile, A., R. A. Morse, and R. Boch. 1975. Swarming honey bees guided by pheromones. *Annals of the Entomological Society of America* 68:1079–1082.

Page 183: For the detailed description of the subtle guide hypothesis, and the results of computer simulations of animal groups making moves using this mechanism of guidance, see Couzin, I. D., J. Krause, N. R. Franks, and S. A. Levin. 2007. Effective leadership and decision making in animal groups on the move. *Nature* 433:513–516.

Page 184: Quote of Martin Lindauer regarding guiding bees flying rapidly through a swarm cloud, from Lindauer, M. 1955. Schwarmbienen auf Wohnungssuche. *Zeitschrift für vergleichende Physiologie* 37:263–324. P. 319. Translated by P. K. Visscher.

Page 184: For the full report of the simulation study of the streaker bee hypothesis, see Janson, S., M. Middendorf, and M. Beekman. 2005. Honeybee swarms: how do scouts guide a swarm of uninformed bees? *Animal Behaviour* 70:349–358.

Pages 184–187: For the full report of the study that tested the hypothesis that scout bees steer a flying swarm using the attraction pheromones produced in their scent organs, see Beekman, M., R. L. Fathke, and T. D. Seeley. 2006. How does an informed minority of scouts guide a honey bee swarm as it flies to its new home? *Animal Behaviour* 71:161–171.

Page 185: For a beautiful description of the anatomy of the scent organ and a review of the chemical composition of the Nasonov gland secretion, see chapter 8, Glands: chemical communication and wax production, in Goodman, L. J. 2003. *Form and Function in the Honey Bee*. International Bee Research Association, Cardiff.

Page 187: For the full report of the check for streaker bees using still photography, see Beekman, M., R. L. Fathke, and T. D. Seeley. 2006. How does an informed minority of scouts guide a honey bee swarm as it flies to its new home? *Animal Behaviour* 71:161–171.

Page 189: Quote of Kevin Passino on cooperative control strategies, from Passino, K. M. 2005. *Biomimicry for Optimization, Control, and Automation*. Springer Verlag, London. P. 80.

Pages 190–193: For the full report of the study in which individual bees in flying swarms were tracked using video analysis, see Schultz, K. M., K. M. Passino, and T. D. Seeley. 2008. The mechanism of flight guidance in honeybee swarms: subtle guides or streaker bees? *Journal of Experimental Biology* 211:3287–3295.

Pages 193–195: For the full report of the experimental test of the streaker bee hypothesis, see Latty, T., M. Duncan, and M. Beekman. 2009. High bee traffic disrupts transfer of directional information in flying honeybee swarms. *Animal Behaviour* 78:117–121.

Chapter 9. Swarm as Cognitive Entity

Page 198: Quote of William Newsome, from the introduction to his talk delivered at the Systems Biology of Decision Making workshop, Mathematical Biosciences Institute, Ohio State University, June 17, 2008.

Page 199: The view that honeybee and other social insect colonies are elegant information-processing devices, and that there are parallels between decision making in social insect colonies and primate brains, has been developed recently in Passino, K. M., T. D. Seeley, and P. K. Visscher. 2008. Swarm cognition in honey bees. *Behavioral Ecology and Sociobiology* 62:401–414; Couzin, I. D. 2008. Collective cognition in animal groups. *Trends in Cognitive Sciences* 13:36–42; and Marshall, J.A.R., R. Bogacz, A. Dornhaus, R. Planqué, T. Kovacs, and

N. R. Franks. 2009. On optimal decision making in brains and social insect colonies. *Journal of the Royal Society Interface* 6:1065–1074.

Pages 199–203: For detailed reviews of the neural basis of primate decision making, see Schall, J. D. 2001. Neural basis of deciding, choosing, and acting. *Nature Reviews Neuroscience* 2:33–42; Glimcher, P. W. 2003. The neurobiology of visual-saccadic decision making. *Annual Review of Neuroscience* 26:133–179; Glimcher, P. W. 2003. *Decisions, Uncertainty, and the Brain: The Science of Neuroeconomics.* MIT Press, Cambridge, MA; Gold, J. I., and M. N. Shadlen. 2007. The neural basis of decision making. *Annual Review of Neuroscience* 30:535–574; and Heekeren, H. R., S. Marrett, and L. G. Ungerleider. 2008. The neural systems that mediate human perceptual decision making. *Nature Reviews Neuroscience* 9:467–479.

Pages 203-204: The Sugrue-Corrado-Newsome framework for thinking about the stages of information processing in making decisions is presented in Sugrue, L. P., G. S. Corrado, and W. T. Newsome. 2005. Choosing the greater of two goods: neural currencies for valuation and decision making. *Nature Reviews Neuroscience* 6:363–375.

Page 210: The Usher-McClelland model of decision making in the primate visual cortex is described in Usher, M., and J. L. McClelland. 2001. The time course of perceptual choice: the leaky, competing accumulator model. *Psychological Review* 108:550–592. An earlier connectionist model of decision making, which is likewise based on the idea that information is sequentially sampled and accumulated over time to make a decision, is described in Busemeyer, J. R., and J. T. Townsend. 1993. Decision field theory: a dynamic cognition approach to decision making. *Psychological Review* 100:432–459. For an excellent general review of the decision models that have been developed by mathematical psychologists, see Smith, P. L., and R. Ratcliff. 2004. Psychology and neurobiology of simple decisions. *Trends in Neurosciences* 27:161–168.

Pages 210–213: For the full report of the mathematical modeling of the nest-site selection by honeybee swarms, which includes the analysis of how the dance decay rate and the quorum size have been tuned by natural selection to achieve a good balance between speed and accuracy in a swarm's decision making, see Passino, K. M., and T. D. Seeley. 2006. Modeling and analysis of nest-site selection by honeybee swarms: the speed and accuracy tradeoff. *Behavioral Ecology and Sociobiology* 59:427–442.

Page 213: See Hofstadter, D. R. 1979. *Gödel, Escher, Bach: An Eternal Golden Braid*. Basic Books, New York.

Page 215: For overviews of the analysis of collective nest choice by colonies of rock ants, see Mallon, E. B., S. C. Pratt, and N. R. Franks. 2001. Individual and collective decision making during nest site selection by the ant *Leptothorax albipennis*. *Behavioral Ecology and Sociobiology* 50:352–359; Franks, N. R., S. C. Pratt, E. B. Mallon, N. F. Britton, and D. J. T. Sumpter. 2002. Information flow, opinion polling and collective intelligence in house-hunting social insects. *Philosophical Transactions of the Royal Society of London B* 357:1567–1583; Pratt, S. C., D. J.T. Sumpter, E. B. Mallon, and N. R. Franks. 2005. An agent-based model of collective nest choice by the ant *Temnothorax albipennis*. *Animal Behaviour* 70:1023–1036; and Franks, N. R., F.-X. Dechaume-Moncharmont, E. Hanmore, and J. K. Reynolds. 2009. Speed versus accuracy in decision-making ants: expediting politics and policy implementation. *Philosophical Transactions of the Royal Society B* 364:845–852.

Pages 215–217: For the full analysis of how social insect colonies may be able to achieve statistically optimal collective decision making in a way similar to primate brains via competition between populations of evidence-accumulating subunits (workers or neurons), see Marshall, J.A.R., R. Bogacz, A. Dornhaus, R. Planqué, T. Kovacs, and N. R. Franks. 2009. On optimal decision making in brains and social insect colonies. *Journal of the Royal Society Interface* 6:1065–1074. This paper rests on a foundation of theoretical studies of optimal decision making. See, for example, Bogacz, R., E. Brown, J. Moehlis, P. Holmes, and J. D. Cohen. 2006. The physics of optimal decision making: a formal analysis of models of performance in two-alternative force choice tasks. *Psychological Review* 113:700–765.

Chapter 10. Swarm Smarts

Page 218: Quote of William Shakespeare, from Shakespeare, W. 1599. *Henry V*. Act I, scene 2, lines 190–192.

Page 218: For general discussions of how a group of humans working face to face can be organized so that the many are reliably smarter than the few, see Elster, J. 2000. *Deliberative Democracy*. Cambridge University Press, Cambridge; Surowiecki, J. 2004. *The Wisdom of Crowds*. Doubleday, New York; and Austen-Smith, D., and T. J. Feddersen. 2009. Information aggregation and communication in committees. *Philosophical Transactions of the Royal Society B* 364:763–769.

Page 218: For authoritative reviews on the fossil record of honeybees (genus *Apis*), see Engel, M. S. 1998. Fossil honey bees and evolution in the genus *Apis* (Hymenoptera: Apidae). *Apidologie* 29:265–281; Engel, M. S. 1999. The taxonomy of recent and fossil honey bees (Hymenoptera: Apidae: *Apis*). *Journal of Hymenoptera Research* 8:165–196; Engel, M. S. 2006. A giant honey bee from the middle Miocene of Japan (Hymenoptera: Apidae). *Journal of the Kansas Entomological Society* 76:71; and Engel, M. S., I. A. Hinojosa-Diaz, and A. Rasnitsyn. 2009. A honey bee from the Miocene of Nevada and the biogeography of *Apis* (Hymenoptera: Apidae: Apini). *Proceedings of the California Academy of Sciences* 60:23–38.

Page 219: The expression "The Five Habits of Highly Effective Groups" is inspired by the title of Stephen R. Covey's excellent book, *The Seven Habits of Highly Effective People*. 1989. Free Press, New York.

Page 219: The New England town meeting is a law-making legislative assembly in which every participating citizen (registered voter) is a legislator. It should not be confused with the "town hall meeting" that has become a popular form of public hearing but is not an assembly that has the force of law. For information on how the New England town meeting works, see Gould, J. 1940. *New England Town Meeting: Safeguard of Democracy.* Stephen Daye Press, Brattleboro, VT; Mansbridge, J. J. 1980. *Beyond Adversary Democracy*. University of Chicago Press, Chicago; and Bryan, F. M. 2004. *Real Democracy: The New England Town Meeting and How It Works.* University of Chicago Press, Chicago.

Page 222: The importance of avoiding leadership practices that bias the group's decision making and foster concurrence-seeking is discussed in detail in Janis, I. L. 1982. *Groupthink*. 2nd ed. Houghton Mifflin, Boston.

Page 223: For a description of the leadership style of President George W. Bush (instinctive rather than intellectual) and how he and his foreign policy team decided to invade Iraq without undertaking an open inquiry and critical evaluation of all possible options, see McClellan, S. 2008. *What Happened*. Public Affairs, New York. Pp. 126–129.

Page 223: For the latest version of the rules of parliamentary law first published in 1876 by Major Henry M. Robert, see Robert, H. M., and S. C. Robert. 2000. *Robert's Rules of Order Newly Revised, 10th Edition*. Perseus Publishing, Philadelphia.

Page 224: For a broad review of how a group of individuals can find better solutions to problems than can a brilliant individual working alone, based in large measure on the group's superior ability to explore diverse options, see Page, S. E. 2007. *The Difference.* Princeton University Press, Princeton, NJ.

Page 226: For an overview of human voting systems and a discussion of the merits of various decision procedures, see Black, D. 1986. *The Theory of Committees and Elections*. Kluwer, Dordrecht.

Page 226: For reviews of the literature on democratic decision making in groups of nonhuman animals, see Conradt, L., and T. J. Roper. 2003. Group decision making in animals. *Nature* 421:155–158; Conradt, L., and T. J. Roper. 2005. Consensus decision making in animals. *Trends in Ecology and Evolution* 20:449–456; Conradt, L. and T. J. Roper. 2007. Democracy in animals: the evolution of shared group decisions. *Proceedings of the Royal Society of London B* 274:2317–2326; Conradt, L., and C. List. 2009. Introduction. Group decisions in humans and animals: a survey. *Philosophical Transactions of the Royal Society B* 364:719–742.

Pages 226–228: A recent modeling study shows explicitly how the bees' collective decision-making system depends on both interdependence and independence among the scout bees. Without interdependence (by sharing information about sites with dances), there is no cascading of interest on the best site. Without independence (in assessing and then advertising sites), there is a cascading of interest but not necessarily on the best site. For details, see List, C., C. Elsholtz, and T. D. Seeley. 2009. Independence and interdependence in collective decision making: an agent-based model of nest-site choice by honeybee swarms. *Philosophical Transactions of the Royal Society B* 364:755–762.

Page 228: For more examples of the danger of information cascades, when decision makers blindly copy the decisions others, see Shiller, R. J. 2000. *Irrational Exuberance*. Princeton University Press, Princeton, NJ; and Thaler, R. H., and C. R. Sunstein. 2008. *Nudge*. Yale University Press, New Haven, CT. Two important articles on the topic are Bikhchandani, S., D. Hirshleifer, and I. Welch. 1992. A theory of fads, fashions, custom, and cultural change as informational cascades. *Journal of Political Economy* 100:992–1026; and Bikhchandani, S., D. Hirshleifer, and I. Welch. 1998. Learning from the behavior of others: conformity, fads, and informational cascades. *Journal of Economic Perspectives* 12:151–170.

Pages 230–231: For a general discussion of the utility of quorum responses in building consensus decisions (i.e., when group members come to agree on the same option), see Sumpter, D. J. T., and S. C. Pratt. 2009. Quorum responses and consensus decision making. *Philosophical Transactions of the Royal Society B* 364:743–753.

Epilogue

Page 233: Quote of Martin Lindauer regarding beautiful experience, from Seeley, T. D., S. Kühnholz, and R. H. Seeley. 2002. An early chapter in behavioral physiology and sociobiology: the science of Martin Lindauer. *Journal of Comparative Physiology A* 188:439–453. P. 447.

Page 233: Unfortunately, it is difficult to apply the honeybees' lessons about good democratic decision making to groups composed of individuals with strongly conflicting interests. In such adversarial groups, individuals will not behave like scout bees: totally honest and reliably hardworking. They are instead expected to issue lies and act lazily when doing so provides them with benefits even if doing so degrades the group's success. Nevertheless, because many small democratic organizations are composed of people with strongly overlapping interests, I feel the lessons learned from the house-hunting bees have considerable relevance to human affairs.

Acknowledgments

Thanks in greatest measure go to my wife, Robin, who I met on Appledore Island, for sharing the adventure of studying bees on that windblown island. Her sharp Yankee wit, love of working six miles out in the North Atlantic Ocean, and steady support over the past 34 years has brought many joys to the explorations described in this book.

Thanks are due also to our daughters, Saren and Maira, for putting up with their father's love affair with honeybees and for sharing times at our remote camp beside Ox Cove, in far-off Pembroke, Maine. I am particularly grateful to their feedback on my writing and for suggesting the title *Honeybee Democracy*.

Much of the work described in this book could not have been done without the help of the graduate students, undergraduate research students, and summer field assistants who have enlivened my laboratory at Cornell over the past quarter century. I have made it a point to acknowledge throughout the book all the students who have investigated various mysteries of the swarm bees, but I would like to also acknowledge here all the research assistants who provided the extra sets of eyes, ears, hands, and brains (!) needed for many of the studies: Susannah Buhrman-Deever, Siobhan Cully, Robert Fathke, Madeleine Girard, Sean Griffin, Benjamin Land, Sasha Mikheyev, Marielle Newsome, Kriston Pastor, Adrian Reich, and Ethan Wolfson-Seeley.

Essential, too, has been the close partnerships that I have enjoyed with my collaborators: Madeleine Beekman, the late Roger A. Morse, Kevin Passino, Jürgen Tautz, and Kirk Visscher. Without their enthusiasm in helping push the work forward, the natural history investigations, the agent-based modeling studies, the video analyses, and many of the experimental investigations reported here would not have been accomplished. The warm friendship of these five people has also added immensely to the pleasure of working together to figure out how swarms work.

Special thanks are due to my earliest mentors—Bernd Heinrich, Bert Hölldobler, the late Martin Lindauer, the late Roger A. Morse, and Edward O. Wilson—for sharing their scientific wisdom and for giving me encouragement when I started studying the bees in earnest back in the mid-1970s. I am profoundly grateful too for their enduring friendship, guidance, and support. Hopefully, each will see strong signs of their influence in this book. Its title, for example, pays homage to Bernd Heinrich's wonderful book *Bumblebee Economics*.

I am also grateful to a number of fellow scientists whose work and friendship has been a source of knowledge and inspiration over the years: Kraig Adler, Andrew Bass, Koos Biesmeijer, Nicholas Britton, Nicholas Calderone, Scott Camazine, Larissa Conradt, Iain Couzin, Brian Danforth, Fred Dyer, the late George Eickwort, Michael Engel, Tom Eisner, Joseph Fetcho, Nigel Franks, Ronald Hoy, Barrett Klein, Susannah Kühnholz, Egbert Leigh, Christian List, James Marshall, Heather Mattila, Matthew Meselson, Randolf Menzel, Sasha Mikheyev, Mary Myerscough, Jun Nakamura, Francis Ratnieks, Kern Reeve, Gene Robinson, Paul Sherman, David Tarpy, Craig Tovey, Walter Tschinkel, Rüdiger Wehner, Anja Weidenmüller, and David S. Wilson.

It has also been my good fortune to get to know over the past decade several remarkable people outside the world of biology. Frank Bryan, professor of political science at the University of Vermont and world authority on New England town meetings, has taught me much about his specialty and introduced me to Larry Coffin, long-standing moderator of the annual town meeting in Bradford, Vermont. I am most grateful to Mr. Coffin and the citizens of Bradford for letting me observe real democracy on Town Meeting Day. Paul Hyams, professor of medieval history at Cornell University, has guided me in studying the history of human democracies and has helped me apply sensibly what I've learned from the bees to human affairs. Michael Mauboussin, chief investment strategist at Legg Mason Capital Management, has showed me the connections between the search committees of bees and the investment committees of humans and kindly allowed me to borrow from one of his Consilient Observer essays the title for my final chapter, "Swarm Smarts." John Miller, professor of social and decision sciences at

Carnegie Mellon University, has kindly linked me to the Santa Fe Institute. This remarkable research establishment has fostered the study of collective decision making by supporting workshops that have brought together people from diverse fields ranging from engineering and economics to neurobiology and behavior. I am deeply fortunate to have been pulled "off course" by these nonbiologists and so exposed to many beautiful ideas related to understanding how a system of leaderless agents can create a collective intelligence.

My 24 years at Cornell have been happy ones, in large measure because of the many personal ties forged over the years. To my fellow Cornellians, including the staff and my colleagues in the Department of Neurobiology and Behavior, I extend my thanks for brightening the cloudy days for which Ithaca is so famous. I am immeasurably indebted to John M. Kingsbury, visionary founder of the Shoals Marine Laboratory, who recognized early on the value of this rugged island for a biological field station devoted to undergraduate education, and who welcomed me (and my bees) to it in 1975. My indebtedness extends to the past directors of the Shoals Marine Laboratory, John Heiser and Jim Morin, as well as to the present director, William Bemis, for making the laboratory a paradise for exploring nature, both on the land and in the sea.

I am always appreciative of the financial support received through the years from the Alexander von Humboldt Foundation, the American Philosophical Society, National Geographic Foundation, the National Science Foundation, and the U.S. Department of Agriculture.

Over the years I have worked with various editors who have helped me present my findings about the bees to a broad audience: Dennis Flanagan and Jonathan Piel (*Scientific American*, 1981 and 1985); Linda Peterson, Don Cunningham, and Fenella Saunders (*American Scientist*, 1982, 1989, and 2006); Robert Wright (*The Sciences*, 1987); Rebecca Finnel (*Natural History*, 2002); Kim Flottum (*Bee Culture*, 1998–2009); and Silke Beckedorf (*Deutsches Bienen Journal*, 2009). I most grateful for their encouragement and advice about conveying to a general reader what it feels like to do science: the sights and sounds, the tedium of data collection, the happy accidents, the obsession born of curiosity, and the sheer joy of discovery.

I am grateful to a number of people, including Bernd Heinrich, Paul

Hyams, James Marshall, Michael Mauboussin, Francis Ratnieks, Kevin Schultz, and Maira Seeley, for commenting on parts of the manuscript. Two close collaborators, Kevin Passino and Kirk Visscher, were kind enough to critique the entire text. I am greatly indebted to Scott Camazine, Marco Kleinhenz, and Rosemarie Lindauer for providing photos.

I also wish to acknowledge how pleasurable it has been to work with Margaret C. Nelson, who created all the illustrations for this book. Without her talent for depicting the bees' behaviors in precise line drawings and her skill in rendering my rough paper sketches into exact computer images, this book would lack much of its visual appeal. I am much indebted to several individuals at Princeton University Press: Alison Kalett, editor for biology and earth science, for steady support and superb guidance all along; Stefani Wexler for help in pulling everything together; and Carmina Alvarez and Heath Renfroe, respectively, for designing the book and for guiding it through the production process. Dawn Hall edited the entire manuscript, displaying at all times the diligence and skill for which she is famous.

To all, I give heartfelt thanks.

Illustration Credits

Fig. 1.1. Photo by Kenneth Lorenzen.

Fig. 1.2. Photo by Thomas D. Seeley.

Fig. 1.3. Modified from Fig. 46 in Frisch, K. von. 1993. *The Dance Language and Orientation of Bees*. Harvard University Press, Cambridge, MA.

Fig. 1.4. Original drawing by Margaret C. Nelson.

Fig. 1.5. Photo provided by Rosemarie Lindauer.

Fig. 1.6. Modified from Fig. 3 in Lindauer, M. 1951. Bienentänze in der Schwarmtraube. *Die Naturwissenschaften* 38:509–513.

Fig. 1.7. Photo by John G. Seeley.

Fig. 2.1. Photo by Kenneth Lorenzen.

Fig. 2.2. Photo by Kenneth Lorenzen.

Fig. 2.3. Original drawing by Margaret C. Nelson.

Fig. 2.4. Photo by Scott Camazine.

Fig. 2.5. Modified from Fig. 54 in Kemper, H., and E. Döhring. 1967. *Die Sozialen Faltenwespen Mitteleuropas*. Paul Parey, Berlin.

Fig. 2.6. Modified from Fig. 8.2 in Seeley, T. D. 1985. *Honeybee Ecology*. Princeton University Press, Princeton, NJ.

Fig. 2.7. Photo by Thomas D. Seeley.

Fig. 2.8. Modified from Fig. 1 in Seeley, T. D., and P. K. Visscher. 1985. Survival of honeybees in cold climates: the critical timing of colony growth and reproduction. *Ecological Entomology* 10:81–88.

Fig. 2.9. Photo by Kenneth Lorenzen.

Fig. 2.10. Photo by Kenneth Lorenzen.

Fig. 2.11. Original drawing by Margaret C. Nelson.

Fig. 2.12. Photo by Kenneth Lorenzen.

Fig. 2.13. Modified from Fig. 3.7 in Seeley, T. D. 1985. *Honeybee Ecology*. Princeton University Press, Princeton, NJ.

Fig. 2.14. Modified from Fig. 2 in Michelsen, A., W. H. Kirchner, B. B. Andersen, and M. Lindauer. 1986. The tooting and quacking vibration signals of honeybee queens: a quantitative analysis. *Journal of Comparative Physiology A* 158:605–611.

Fig. 3.1. Reproduction of Fig. 20.3a in Crane, E. 1999. *The World History of Beekeeping and Honey Hunting*. Routledge, New York.

Fig. 3.2. Photo by Thomas D. Seeley.

Fig. 3.3. Photo by Thomas D. Seeley.

Fig. 3.4. Modified from Fig. 2 in Seeley, T. D., and R. A. Morse. 1976. The nest of the honey bee (*Apis mellifera* L.). *Insectes Sociaux* 23:495–512.

Fig. 3.5. Photo by Thomas D. Seeley.

Fig. 3.6. Photo by Thomas D. Seeley.

Fig. 3.7. Photo by Thomas D. Seeley.

Fig. 3.8. Modified from Fig. 3 in Morse, R. A., and T. D. Seeley. Bait hives. *Gleanings in Bee Culture* 106 (May):218–220, 242.

Fig. 3.9. Photo by Thomas D. Seeley.

Fig. 3.10. Aerial photo courtesy of the Shoals Marine Laboratory; landscape photo by Thomas D. Seeley.

Fig. 3.11. Photo by Thomas D. Seeley.

Fig. 3.12. Photo by Thomas D. Seeley.

Fig. 3.13. Modified from Fig. 2 in Seeley, T. D. 1982. How honeybees find a home. *Scientific American* 247 (October):158–168.

Fig. 3.14. Modified from Fig. 8 in Seeley, T. D. 1982. How honeybees find a home. *Scientific American* 247 (October):158–168.

Fig. 4.1. Photo by Peter Essick.

Fig. 4.2. Modified from Fig. 3 in Lindauer, M. 1955. Schwarmbienen auf Wohnungssuche. *Zeitschrift für vergleichende Physiologie* 37:263–324.

Fig. 4.3. Modified from Fig. 4 in Lindauer, M. 1955. Schwarmbienen auf Wohnungssuche. *Zeitschrift für vergleichende Physiologie* 37:263–324.

Fig. 4.4. Modified from Fig. 7 in Lindauer, M. 1955. Schwarmbienen auf Wohnungssuche. *Zeitschrift für vergleichende Physiologie* 37:263–324.

Fig. 4.5. Photo by Thomas D. Seeley.

Fig. 4.6. Modified from Fig. 3 in Seeley, T. D., and S. C. Buhrman. 1999. Group decision making in swarms of honey bees. *Behavioral Ecology and Sociobiology* 45:19–31.

Fig. 4.7. Modified from Fig. 5 in Seeley, T. D., and S. C. Buhrman. 1999. Group decision making in swarms of honey bees. *Behavioral Ecology and Sociobiology* 45:19–31.

Fig. 4.8. Reproduction of Fig. 16 in Lindauer, M. 1955. Schwarmbienen auf Wohnungssuche. *Zeitschrift für vergleichende Physiologie* 37:263–324.

Fig. 4.9. Modified from Fig. 2 in Gilley, D. C. 1998. The identity of nest-site scouts in honey bee swarms. *Apidologie* 29:229–240.

Fig. 5.1. Modified from Fig. 1 in Seeley, T. D., and S. C. Buhrman. 2001. Nest-site selection in honey bees: how well do swarms implement the "Best-of-N" decision rule? *Behavioral Ecology and Sociobiology* 49:416–427.

Fig. 5.2. Photo by Thomas D. Seeley.

Fig. 5.3. Modified from Fig. 2 in Seeley, T. D., and S. C. Buhrman. 2001. Nest-site selection in honey bees: how well do swarms implement the "Best-of-N" decision rule? *Behavioral Ecology and Sociobiology* 49:416–427.

Fig. 5.4. Photo by Thomas D. Seeley.

Fig. 5.5. Modified from Fig. 4 in Seeley, T. D., and S. C. Buhrman. 2001. Nest-site selection in honey bees: how well do swarms implement the "Best-of-N" decision rule? *Behavioral Ecology and Sociobiology* 49:416–427.

Fig. 5.6. Photo by Thomas D. Seeley.

Fig. 5.7. Modified from Fig. 5 in Seeley, T. D., and S. C. Buhrman. 2001. Nest-site selection in honey bees: how well do swarms implement the "Best-of-N" decision rule? *Behavioral Ecology and Sociobiology* 49:416–427.

Fig. 6.1. Original drawing by Margaret C. Nelson.

Fig. 6.2. Original drawing by Margaret C. Nelson.

Fig. 6.3. Photo by Thomas D. Seeley.

Fig. 6.4. Photo by Thomas D. Seeley.

Fig. 6.5. Modified from Fig. 2 in Seeley, T. D., and P. K. Visscher. 2008. Sensory coding of nest-site value in honeybee swarms. *Journal of Experimental Biology* 211:3691–3697.

Fig. 6.6. Modified from Fig. 3 in Seeley, T. D., and P. K. Visscher. 2008. Sensory coding of nest-site value in honeybee swarms. *Journal of Experimental Biology* 211:3691–3697.

Fig. 6.7. Modified from Fig. 11 in Seeley, T. D., P. K. Visscher, and K. M. Passino. 2006. Group decision making in honey bee swarms. *American Scientist* 94:220–229.

Fig. 6.8. Modified from Fig. 18 in Lindauer, M. 1955. Schwarmbienen auf Wohnungssuche. *Zeitschrift für vergleichende Physiologie* 37:263–324.

Fig. 6.9. Modified from Fig. 1 in Seeley, T. D. 2003. Consensus building during nest-site selection in honey bee swarms: the expiration of dissent. *Behavioral Ecology and Sociobiology* 53:417–424.

Fig. 6.10. Modified from Fig. 2 in Seeley, T. D. 2003. Consensus building during nest-site selection in honey bee swarms: the expiration of dissent. *Behavioral Ecology and Sociobiology* 53:417–424.

Fig. 7.1. Photo by Peter Essick.

Fig. 7.2. Photos by Thomas D. Seeley.

Fig. 7.3. Modified from Fig. 23 in Heinrich, B. 1981. The mechanisms and energetics of honeybee swarm temperature regulation. *Journal of Experimental Biology* 91:25–55.

Fig. 7.4. Photo by Marco Kleinhenz and Thomas D. Seeley.

Fig. 7.5. Modified from Fig. 2 in Seeley, T. D., M. Kleinhenz, B. Bujok, and J. Tautz. 2003. Thorough warm-up before take-off in honey bee swarms. *Naturwissenschaften* 90:256–260.

Fig. 7.6. Modified from Fig. 2 in Seeley, T. D., and J. Tautz. 2001. Worker piping in honey bee swarms and its role in preparing for liftoff. *Journal of Comparative Physiology A* 187:667–676.

Fig. 7.7. Modified from Fig. 8 in Seeley, T. D., and J. Tautz. 2001. Worker piping in honey bee swarms and its role in preparing for liftoff. *Journal of Comparative Physiology A* 187:667–676.

Fig. 7.8. Modified from Fig. 4 in Seeley, T. D. and J. Tautz. 2001. Worker piping in honey bee swarms and its role in preparing for liftoff. *Journal of Comparative Physiology A* 187:667–676.

Fig. 7.9. Modified from Fig. 7 in Seeley, T. D., and J. Tautz. 2001. Worker piping in honey bee swarms and its role in preparing for liftoff. *Journal of Comparative Physiology A* 187:667–676.

Fig. 7.10. Modified from Fig. 6.3 in Seeley, T. D. 1995. *The Wisdom of the Hive.* Harvard University Press, Cambridge, MA.

Fig. 7.11. Modified from Fig. 1 and Fig. 9 in Seeley, T. D., and J. Tautz. 2001. Worker piping in honey bee swarms and its role in preparing for liftoff. *Journal of Comparative Physiology A* 187:667–676.

Fig. 7.12. Modified from Fig. 1 in Rittschof, C. C., and T. D. Seeley. 2008. The buzz-run: how honeybees signal "Time to go!" *Animal Behaviour* 75:189–197.

Fig. 7.13. Modified from Fig. 2 and Fig. 3 in Rittschof, C. C., and T. D. Seeley. 2008. The buzz-run: how honeybees signal "Time to go!" *Animal Behaviour* 75:189–197.

Fig. 7.14. Photo by Thomas D. Seeley.

Fig. 8.1. Photo by Peter Essick.

Fig. 8.2. Photo by Thomas D. Seeley.

Fig. 8.3. Modified from Fig. 2 in Beekman, M., R. L. Fathke, and T. D. Seeley. 2006. How does an informed minority of scouts guide a honeybee swarm as it flies to its new home? *Animal Behaviour* 71:161–171.

Fig. 8.4. Photo by Kenneth Lorenzen.

Fig. 8.5. Modified from Fig. 42 in Frisch, K. von. 1967. *The Dance Language and Orientation of Bees.* Harvard University Press, Cambridge, MA.

Fig. 8.6. Photo by Thomas D. Seeley.

Fig. 8.7. Modified from Fig. 4 in Schultz, K. M., K. M. Passino, and T. D. Seeley. 2008. The mechanism of flight guidance in honeybee swarms: subtle guides or streaker bees? *Journal of Experimental Biology* 211:3287–3295.

Fig. 8.8. Original drawing by Margaret C. Nelson.

Fig. 8.9. Modified from Fig. 1 in Latty, T., M. Duncan, and M. Beekman. 2009. High bee traffic disrupts transfer of directional information in flying honeybee swarms. *Animal Behaviour* 78:117–121.

Fig. 9.1. Original drawing by Margaret C. Nelson.

Fig. 9.2. Modified from Fig. 4 in Glimcher, P. W. 2003. The neurobiology of visual-saccadic decision making. *Annual Review of Neuroscience* 26:133–179.

Fig. 9.3. Modified from Fig. 1 in Seeley, T. D., and P. K. Visscher. 2008. Sensory coding of nest-site value in honeybee swarms. *Journal of Experimental Biology* 211:3691–3697.

Fig. 9.4. Original drawing by Margaret C. Nelson.

Fig. 9.5. Original drawing by Margaret C. Nelson.

Fig. 9.6. Original drawing by Margaret C. Nelson.

Index

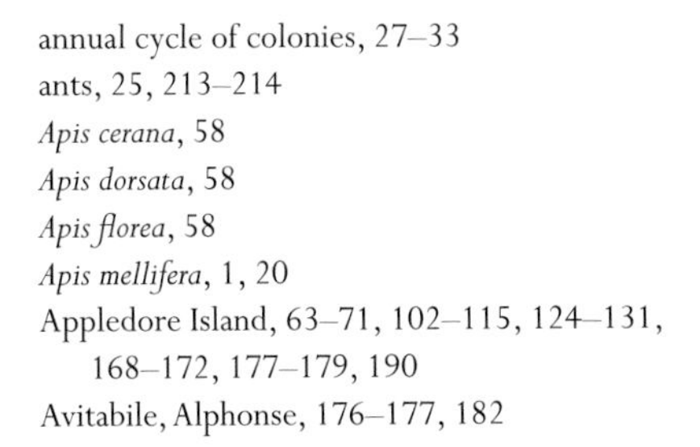